T0145325

Studies in Systems, Decision and Control

Volume 321

Series Editor

Janusz Kacprzyk, Systems Research Institute, Polish Academy of Sciences, Warsaw, Poland

The series "Studies in Systems, Decision and Control" (SSDC) covers both new developments and advances, as well as the state of the art, in the various areas of broadly perceived systems, decision making and control–quickly, up to date and with a high quality. The intent is to cover the theory, applications, and perspectives on the state of the art and future developments relevant to systems, decision making, control, complex processes and related areas, as embedded in the fields of engineering, computer science, physics, economics, social and life sciences, as well as the paradigms and methodologies behind them. The series contains monographs, textbooks, lecture notes and edited volumes in systems, decision making and control spanning the areas of Cyber-Physical Systems, Autonomous Systems, Sensor Networks, Control Systems, Energy Systems, Automotive Systems, Biological Systems, Vehicular Networking and Connected Vehicles, Aerospace Systems, Automation, Manufacturing, Smart Grids, Nonlinear Systems, Power Systems, Robotics, Social Systems, Economic Systems and other. Of particular value to both the contributors and the readership are the short publication timeframe and the world-wide distribution and exposure which enable both a wide and rapid dissemination of research output.

** Indexing: The books of this series are submitted to ISI, SCOPUS, DBLP, Ulrichs, MathSciNet, Current Mathematical Publications, Mathematical Reviews, Zentralblatt Math: MetaPress and Springerlink.

More information about this series at http://www.springer.com/series/13304

Sina Razvarz · Raheleh Jafari · Alexander Gegov

Flow Modelling and Control in Pipeline Systems

A Formal Systematic Approach

Sina Razvarz
Departamento de Control Automatico
CINVESTAV-IPN (National Polytechnic Institute)
Mexico City, Mexico

Raheleh Jafari
School of design
University of Leeds
Leeds, UK

Alexander Gegov
School of Computing
University of Portsmouth
Portsmouth, UK

ISSN 2198-4182 ISSN 2198-4190 (electronic)
Studies in Systems, Decision and Control
ISBN 978-3-030-59248-6 ISBN 978-3-030-59246-2 (eBook)
https://doi.org/10.1007/978-3-030-59246-2

This Springer imprint is published by the registered company Springer Nature Switzerland AG
The registered company address is: Gewerbestrasse 11, 6330 Cham, Switzerland

To my lovely wife and our newborn baby Artin

Sina Razvarz

To my parents, parents in law,
my brothers and specially my husband

For their endless love, support and encouragement

Raheleh Jafari

To the visionaries for the United States of Europe

Alexander Gegov

Preface

A pipeline system as one of the effective tools for transporting fluids despite the cost of proper maintenance, has been taken to be a complex system accompanied by several kinds of components and consumers. Hence, pipeline systems have been taken as one of the most important tools for transmission around the world.

It will be important for industrial society that pipeline systems function appropriately by taking into consideration the growing requirement for effective interconnecting fluid networks. However, this task is difficult as someone should simultaneously certify a secure fluid supply and the fulfillment of the various requirements of consumers. Even this task could become more difficult with the appearance of leakage, blockage, and fault in sensors and actuators that could produce the degradation and glitch of the whole system.

Leakage and blockage in the system of pipes that transport process fluids such as oil, industrial gas, water could result in crucial environmental, social, economic, health and safety problems. Leakage in the pipeline can be caused from poor mechanism or from any devastating reason because of unexpected alterations of pressure, corrosion, fractures, faults in pipelines or absence of preservation. There exist various non-destructive testing (NDT) techniques to detect these faults in pipe networks like radiographic, ultrasonic, magnetic particle inspection, pressure transient and acoustic wave techniques.

The model structure of flow in pipe or pump could be designed by various techniques. One well-known technique is to present flow in pipe using two partial differential equations. In general, the closed-form solution of this method is not known, but it may be obtained based on numerical techniques. Another method of modeling is based on the use of the hydro-electrical analogy.

Over the past few years, various techniques involving uncertainties have been used for detecting flaws in pipelines. Various numerical tests have been carried out for improving the current approaches by taking into consideration parametric studies. The theoretical research focuses on evaluation the precision, robustness, calculational ability, applicability and limitations of the methodology. To achieve the safe operation of pipeline systems, special software tools have been produced in the past decades that are supplementary to the conventional supervisory control and data acquisition systems (SCADA). Generally, those tools are made of fault detection, location and diagnosis algorithms, based on fluid mechanics for signal processing and also, they consider a finite number of existing variables from the pipe.

It should be noted that some defects to be identified need active recognition, for instance, the requirement of supervision systems upon the pipeline system in regular intervals or at acute time by applying test signals for generating, for example, transitory answers of the fluid to detect unusual occurrences. Hence, there exist a great number of research groups throughout the world with various backgrounds who are attempting to develop efficient automated monitoring and supervision systems for pipelines.

The background material needed for understanding this book is fluid dynamic and linear and nonlinear systems. This book will provide a good basis for those students who are interested in numerical analysis and partial differential equations. This book is mainly written for graduate and advanced undergraduate students of sciences, technology, engineering, and mathematics. It is organized as a textbook for a course on control and modeling. This book could be used for self-learning.

In this book we have rather attempted to unify the theory as far as possible with the practice by focusing attention on the most important methods to deal with the general problem. Our aim in this book is to introduce new methods using auxiliary systems called "observers" for solving the defect detection and identification problem in pipe networks and also develop the nonlinear equations for pipe networks. In reading this book, a reader who wants a general knowledge about fluid dynamic and pipeline should read Chaps. 1–5. These chapters provide an understanding of why pipelines are important (Chap. 1), a review on different pipeline fault detection techniques (Chap. 2), mechanisms of fluid flows in pipes (Chap. 3), flow control of fluid in pipelines using fuzzy logic controllers (Chap. 4), flow control of fluid in pipelines using neural networks and deep learning (Chap. 5), model structure of leakage in pipes (Chap. 6). The latter half of the book delves into some introduction to flow control techniques, model structure of blockage in pipes (Chap. 7) leakage detection in pipeline based on observation techniques (Chap. 8) flow control of fluid in pipelines using proportional-derivative (PD) and proportional–integral–derivative (PID) controllers (Chap. 9)

The authors contributed to shape the substance of this book are from computer science and engineering backgrounds. The first author, Sina Razvarz, would like to express his sincere gratitude to his advisor Prof. Cristobal Vargas for his continuous support of his Ph.D. study and research, and for his patience, motivation, enthusiasm, and immense knowledge. His guidance helped him throughout his research and writing of this book. Also, he would like to thank his wife for her time and dedication. Without her this book would not have been possible. The second author, Raheleh Jafari would like to thank her husband for his time and dedication. Without him this book would not have been possible. The third author, Alexander Gegov would like to thank his family members for their spiritual support during the work on this book.

The authors of this book would like to thank the editors for their effective cooperation and great care making possible the publication of this book.

Mexico City, Mexico
Leeds, UK
Portsmouth, UK
July 2020

Sina Razvarz
Raheleh Jafari
Alexander Gegov

Contents

Chapter 1
The Importance of Pipeline Transportation

1.1 Introduction

Over the last few decades, rapid technological advancements or variations in processes have removed or decreased risks related to particular specifications in operations (pipe manufacturing techniques or pipe installation processes). Some of these advancements have happened quickly, therefore pipelines built after the advancement display remarkably improved performance. From the viewpoint of these advancements, one could properly evaluate the particular risk factors associated with the pipeline, and with the information on once the advancements happened, one could describe the effect of the advancement on performance. The combination can create an approach for pipeline operators to evaluate risk factors in their networks and to give priority to mitigation programs.

1.2 History

Two thousand years ago, the ancient Romans used great conduits for transporting water from high altitudes by constructing the conduits in graduated sections that permitted gravity to force the water to move along until it arrived its destination. Hundreds of these systems had been constructed all over Europe and in other places, and along with four mills of the Roman Empire. Furthermore, the past people in China used conduits and pipe networks for public works. The famous Han Dynasty court eunuch Zhang Rang (d. 189 AD) one time commanded the engineer Bi Lan to build a system of square-pallet chain pumps on the regions outside the country's capital city named Luoyang [1]. The constructed chain pumps had been used in imperial palaces as well as living quarters of the Luoyang as the water raised with the chain pumps was imported with an earthenware pipeline system [1].

Pipe systems were constructed more than 5000 years ago by the Egyptians who applied copper pipelines to transfer drinking water to their cities. The primary usage

S. Razvarz et al., *Flow Modelling and Control in Pipeline Systems*, Studies in Systems, Decision and Control 321, https://doi.org/10.1007/978-3-030-59246-2_1

of pipe systems to transfer hydrocarbons goes back to nearly 500 BC in China in which bamboo pipes had been employed to transfer fossil gas for utilization as fuel from drilling holes near the surface of the ground. The fossil gas had been later applied as fuel for boiling saltwater, generating steam that had been condensed into usable drinking water.

Pipe networks are beneficial to transfer water for drinking purpose or irrigation at a great distance whenever it requires to go up and down hills, or where canals or channels are wrong options because of the considerations of vaporization, pollution, or environmental effect. The 530 km (330 mi) Goldfields Water Supply project in Western Australia employing a 750 mm (30 in.) pipeline and ended in 1903 had been taken to be the greatest water supply project at that time [2]. The Snowy Mountains project [3, 4] was another example of water pipe systems in South Australia and had been divided into two parts, the Morgan-Whyalla pipe system [5] (ended 1944) and Mannum-Adelaide pipe system [6] (ended 1955). There exist two projects for a wide usage of pipe systems namely the Owens Valley Channel (ended 1913) and the Second Los Angeles Channel (ended 1970) occurred in Los Angeles, California. 3,680,000 cubic meters of water are daily supplying to Tripoli, Benghazi, Sirte, and many other towns in Libya by the Great Man-Made River located in Libya. The pipe system has more than 2800 km (1700 mi) length, also is attached to wells and gets the water from aquifers above 500 m (1600 ft) belowground [4].

Pipeline transportation is the long-way transfer of a fluid (i.e., liquid, gas, or multiphase) through a pipe system typically from production to consumption. The current year's data from 2014 provides a total of slightly lower than 2,175,000 miles (3,500,000 km) of the pipe system in 120 countries worldwide [7]. 65% of pipe systems had been located in the United States, 8% in Russia, and 3% in Canada, therefore, 75% of all pipe systems had been placed in these three countries [7, 8]. Based on worldwide pipeline construction reports 118,623 miles (190,905 km) of pipe systems are planned and under construction. From this report, 88,976 miles (143,193 km) refer to projects that are in the planning stage and 29,647 miles (47,712 km) refers to projects that are in the construction stage. Fluids like liquids and gases can be transferred in pipe systems and also every stable chemical substance can be transported through pipes [8]. The pipeline is a useful tool for transferring crude and refined petroleum, fuels like alcohol, biodiesel and natural gas, and fluids like hot water or steam for a short distance. It is also an effective tool for transferring water for drinking purposes or irrigation at a great distance whenever it requires going up and down hills, or where canals or channels are wrong options because of the considerations of vaporization, pollution, or environmental effect [7]. It has been stated about 400 BC wax-coated bamboo pipes were employed for bringing fossil gas into towns, illuminating China's capital, Peking. The latter half of the nineteenth century was a period of great change in the structure of pipelines and their growth in size and number.

During drilling for the purpose of extracting water, crude oil was incidentally found in underground reservoirs. In the beginning, the crude oil was not in high demand until simple refineries been created. The crude oil was transferred to the refineries in wooden tanks and were even transferred by barges pulled by horses across rivers. The transferred crude oil was evaporated in refineries to produce the by-products of naphtha, petroleum and benzene. The petroleum was employed as a fuel to produce light and in the beginning, benzene was considered an undesired item and was thrown away. Railway tanker cars were another way to transport crude oil. Nevertheless, large railway owners used to control the oil supply. Therefore, in order to have a cheap and independent transportation, companies began to adopt pipelines as a more efficient and economical tool of transport. The situation altered significantly after the creation of the automobile that immediately enhanced the request for regular and trusty supplies of gasoline and led to the demand for many more pipes. Nowadays pipe systems transfer a great range of materials such as coal, crude oil, gasoline, natural gases, liquid condensate, process gases, and petroleum. Now there exist about 1.2 million miles of pipe systems for transportation throughout the world and also some well with more than 1000 miles longitude. The full length of these pipe systems if laid end-to-end would circle the earth 50 times over.

1.3 Material of Pipeline

Oil pipes are composed of a variety of materials, including steel and plastic tubes and are generally buried underground. The oil moves overall the pipe systems by pump stations along the pipe systems. Fossil gases (also similar gaseous fuels) pressurize into liquids and are called Natural Gas Liquids (NGLs) [9]. The pipes for transporting natural gas are mainly made of carbon steel. Pipelines can be also used for transporting hydrogen. Pipelines as the safest means of transferring materials over long distances in comparison with road or rail are usually the target of army attacks in time of war.

When it comes to the construction of the first crude oil pipeline, there is no specific date [10]. However, there is a disputation on the first initiation of the pipe systems [11] between Vladimir Shukhov and the Branobel company at the end of the nineteenth century, and the Oil Transport Association that for the first time in the 1860s built a 2-inch (51 mm) wrought iron pipe system over a distance of 6-mile (9.7 km) from an oil field in Pennsylvania to a railroad station in Oil Creek. Pipelines have been the most economical way to carry tens of millions of metric tons of oil and gas over lands. For instance, in 2014, the cost of transportation of crude oil using pipe networks was nearly \$5 per barrel, whereas the cost of rail transportation was nearly \$10 to \$15 per barrel [11]. Trucking has even higher transportation costs compared with rail transportation because of the need for labor [11]. In the United States, 70% of crude oil and petroleum are transporting by pipes, 23% by ships, 4% by trucks, and 3% by rails. In Canada, 97% of natural gas and petroleum are transporting by pipes [11].

Natural gas (also similar gaseous fuels) could be turned into liquids called natural gas liquids when it is gently pressurized. Natural gas liquid processing facilities separate a stream of raw natural gas into methane and natural gas liquids such as butane and propane. The butane and propane liquid under light pressure of 125 lb per square inch (860 kPa), could be transferred by rails, trucks or pipes. Propane as a fuel in oil fields is typically employed to heat different facilities utilized by the oil drillers or instrument and trucks utilized in the oil patch. Propane is a compressible gas that turns to liquid under light pressure, 100 psi, give or take based on the temperature, and is pumping into vehicles at less than 125 psi (860 kPa) at propane fueling stations. Pipes and rail cars employ around twice that pressure for pumping at 250 psi (1700 kPa) [12]. Since most of the natural gas processing plants have been placed in or around oil fields, therefore there are very short shipping distances to markets. The majority of Bakken Basin oil companies located in North Dakota, Montana, Manitoba and Saskatchewan gas areas divide the natural gas liquids in the field, permitting the drillers to sell propane straightly to their customers, removing the refinery product and price controls for propane or butane.

One of the current main pipelines is located in North America known as a TransCanada natural gas line that goes north along with the bridge crossing the Niagara river with Marcellus shale gas from Pennsylvania and others tied in methane or natural gas sources, into Ontario one of the biggest provinces in Canada from fall 2012, providing 16% of the total natural gas utilized in Ontario. This new supply of natural gas has been significantly displaced the natural gas previously transferred to Ontario from western Canada in Alberta and Manitoba, hence decreasing the government regulated pipelines transport costs as there is a short shipping distance between the gas source and consumer. For avoiding any delay and also to neglect the United States government rule, many of the companies responsible for the production in North Dakota came to a common conclusion that they set up an oil pipe system north to Canada to grantee the Canadian oil pipeline transportation system from west to east. This permits the Bakken Basin as well as Three Forks companies responsible for oil production to achieve the best price for their products as they don't need to be restricted to only one wholesale market in the United States. The distance between the greatest oil patch located in North Dakota, in Williston, North Dakota and the Canada–United States border and Manitoba is nearly 85 miles or 137 km.

Mutual funds and joint ventures are now commonly used in almost all major industries and are great investors in oil and gas pipe systems. In the fall of 2012, the United States started exporting liquefied petroleum gas (LPG) to Europe, as wholesale fuel prices in Europe are at much higher levels in comparison with North America. Moreover, a pipe system generally known as Dakota Access Pipeline has been recently built from North Dakota to Illinois. The rapid increase in pipeline construction in North America resulted in the increment of the exportation of liquefied natural gas, propane, butane, and other natural gas products on the three coasts of the United States. To explain, the oil production in North Dakota Bakken has been increased by 600% between 2007 and 2015 [13]. Tanker rail car carries a massive amount of oil from North Dakota oil companies to the market which provides the best deal on prices. The rail cars have been utilized for avoiding a clogged oil pipeline to take the oil to another pipeline for taking the oil to market more rapidly. Nevertheless, transporting oil by pipeline is cheaper than rail cars.

Enbridge planned to reverse the flow of oil in its Line 9 pipeline going through Ontario and Quebec [14, 15]. From a presently rated 250,000 barrels of petroleum every day, it can be increased to between 1 million and 1.3 million barrels daily. New flows on Line 9 bring western oil and feed refineries in Ontario, Michigan, Ohio, Pennsylvania, Quebec, and New York. The Enbridge Sandpiper pipe network with 24–30 in. in diameter has been suggested to carry oil from Western North Dakota through northwestern Minnesota such that it transports over 300,000 barrels of oil per day having the volatility of 32 [16]. Crude oil from western Canada can also be refined in New Brunswick and is exporting to Europe from its deep-water oil ultra-large crude carrier loading port.

Although pipelines can be constructed under the ocean, the pipeline building process is a complex and technically challenging one, hence most of the oil transportation is usually done via tanker ships. Likewise, it is economically easy and safe to carry natural gas in the form of liquefied natural gas, nevertheless, the breakeven costs between liquefied natural gas and pipelines rely on the volume of natural gas and the distance transportation of liquefied natural gas goes by pipeline [15]. The oil and gas pipeline construction industry reported enormous growth before the economic crisis in 2008. A year later, the amount of request for pipeline increased the next year as fuel production grew rapidly [17]. By 2012, nearly 32,000 miles of pipe systems in North America were being either planned or built [18]. However, whenever oil production industries are facing pipeline transportation constraints, truck or rail could be good transportation options to transport products.

Oil pipes may be manufactured from steel or plastic tubes where the interior diameter normally varies from 4 to 48 in. (100–1220 mm). Natural gas pipelines carrying natural gas are made of carbon steel and change in size from 2 to 60 in. (51–1524 mm) in diameter, based on the kind of pipe. Liquid petroleum gas and natural gas could be pressurized using compressor stations and are odorless except that blended with a mercaptan odorant in a case that needed by a regulating authority. The majority of pipes are usually buried at a depth of about 3–6 feet (0.91–1.83 m) underground. For protecting pipelines from gouge, abrasion, and penetration, a variety of techniques have been utilized such as wood lagging (wood slats), concrete coating, rock shield, high-density polyethylene, imported sand padding, and padding machines [19].

Crude oil has different quantities of paraffin wax and in a cold climate, the wax buildup can be created inside a pipe. In most cases, these pipes are checked and cleaned utilizing pigging, the practice of employing tools called “pigs” to carry out different preserving operations on a pipe. The tools are furthermore called “scrapers” or “Go-devils". “Smart pigs” (moreover called “intelligent” or “intelligence” pigs) have been employed to find anomalies in the pipeline like dents, metal loss generated by corrosion or cracking [20]. These tools could proceed from pig-launcher centers and move through the pipe system to be captured at other centers down-stream, either cleaning wax deposits and substance that could have gathered inside the pipeline or checking and recording the situation of the pipeline.

In 1998 the petroleum pipe industry started a voluntary reporting initiative, called the “Pipeline Performance Tracking System.” The aims and objectives of the industry in developing the reporting program were to produce a device for improving safety performance utilizing the information toward zero spills. Precise as well as detailed data are two main factors for learning from events hence operations can be altered for prevention and also to track progress for a period of time. This reporting program gathers better data on spills as small as 5 gallons, indicating the initial time that information on such small issues has been gathered in the entire industry. It furthermore gathered novel information on the construction and infrastructure of the industry, composed of petrol consumption by ten-year-old of building, the petrol consumption by state, the petrol consumption by diameter, and further characteristics of the foundation [21].

For dozens of years, the petroleum pipe industry has been decreased risk and cost in operation by progress in pipeline construction technology and alterations in pipe manufacturing practices. Some of this progress has happened over comparatively short periods of time, therefore, pipes built after the progress displayed notably improved performance. Taking advantage of these advances one can easily evaluate the specific threats to pipe, and also by having information on when the advances could happen, one can describe the effect of the advance on performance. The accessibility of the pipeline performance tracking system mileage information demonstrated the first opportunity to study existing, publicly accessible incident data, describing the effect of the progress in technology and practices on efficiency [22].

1.3.1 Steel

The material used for oil and gas pipes includes steel, particularly either low-carbon steel or low-alloy steel. These two kinds of substances are mainly made of iron (98–99%), very tiny quantities of carbon (0.001–0.30% by weight), manganese (0.30–1.50% by weight), also some other deliberately added alloying components in tiny quantities such as columbium, molybdenum, vanadium, and titanium could have useful impacts on the stability and the fracture hardness of steel. The fracture hardness is the capability of the materials to resist crack propagation [23]. Low-carbon or low alloy steels are relatively inexpensive and appropriate for production of pipe also many steel constructions apply to all building types and bridges as they supply a long-lasting, solid material for withstanding the service loads imposed on such constructions. Other iron-based alloys like wrought iron consists of almost pure iron and cast iron usually with a high carbon content are either too weak or too breakable to perform well as building materials. Rustproof or high-material steels are necessary for particular usages like in high-temperature service and pressure vessels or tool steels. However, they are not appropriate and cannot be manufactured economically in the amounts required for utilization in structures like pipelines. Low-carbon steels or low-alloy steels have good enough strength, toughness, ductility, and weldability for building frames in the construction industry. Line-pipe steels are a category of low-carbon or low-alloy steels and are cost-effective and durable materials. Within a specified temperature range in which these materials are generally used [24] (−20 to +250°F), their characteristics and stability do not change through time. Tensile tests or hardness tests performed nowadays on a low-carbon-steel, one of the most useful materials in the industry, made in 1910 will lead to similar outcomes as tests that could have been performed on the same material back in 1910.

Low-carbon and low-alloy steels are highly sensitive to corrosion in natural environments such as air, water, and soil. However, appropriate coating materials and the usage of a suitable amount of electrical direct current often referred to as cathodic protection could provide satisfactory corrosion protection. Corrosion takes place on the surface of the iron due to the formation of electrochemical cells and causes the iron to become oxidized. Iron oxides are weak and breakable and do not have the ability to carry the loads that are considerably produced by the steel construction. Therefore, corrosion may decrease the strength of a low-carbon or low-alloy steel construction like a pipeline. Providing a sufficient quantity of cathodic protection to an exposed steel surface, can reduce the loss of electrons and slow down the corrosion to a negligible amount. A coat of a protective material applied to steel can successfully prevent corrosion by removing exposed surfaces. Periodic pipe-to-soil potential surveys could be utilized to maintain the cathodic protection of the pipe. Therefore, the risk of corrosion is insignificant in pipelines that are sufficiently coated and cathodically protected [25].

1.3.2 Stress Cycles

Low-carbon and low-alloy steels could survive the infinite number of load/unload cycles in the ranges of stress. When there are a fault or defect, frequent loadings, normally many thousands of cycles could generate fatigue crack development that can cause eventual failure of the construction. However, in-service inspections can be used to detect the presence, location, and size of different kinds of defects that should be immediately repaired before they cause fatal or in-service operations failures [25].

1.3.3 Manufacture and Fabrication

Manufacturing processes for line-pipe steels as well as line pipes have changed significantly over the past century, resulting in significant increases in strength, toughness, ductility, and weldability. Current strategies to improve materials manufacturing processes and quality control measures have resulted in the production of very few manufacturing flaws. Pipeline hydrostatic pressure testing has been used since the late 1960s of the newly made pipeline to present that the pipeline is fit for service. Moreover, by rules, pipelines shall be abandoned or taken temporarily out of service if they have not been tested at the moment of installation hence pressure tightness can be tested. Pipeline hydrostatic pressure testing is a damaging test representing flaws that cause failure at the pressure testing or demonstrates that those defects staying after the test are very small for withstanding the maximum pressure capacity [26].

1.3.4 *Inspection and Testing*

Nondestructive testing that employs sensor and imaging technology and "in-line" inspection tools have been used for more than four decades to recognize irregularities and defects in the pipeline or the coating. Before delivery, the operator monitors the coating for any problems and damages that could have happened during the installation process. The initial in-line inspection device was called an inspection pig or a smart pig and constructed in the mid-1960s. Smart pigs are devices that travel through the pipeline by fluid flow. The devices record cracks and flaws utilizing either ultrasonic wall thickness evaluations or magnetic field disturbances [27]. Advances in technologies led to the design of instrumented pigs smarter, the best way to access and interpret data. The position of a pipeline wall anomaly could be marked by a global positioning system (GPS), showing where increased coatings or even repair would be justified. Further enhanced devices now can recognize specialized defects like fatigue cracks, dents, and mechanical damages [28].

1.4 Implications for Pipeline Safety

The key issues regarding the strength and stability of steel as they impact the safety of pipelines are as follows [29]. The first criteria are that the steel itself does not deteriorate over a long time. With proper protection eighty-year-old pipe displays similar characteristics if analyzed at present as it would have if analyzed 80 years ago.

New construction technologies developed in recent years confirm that a line-pipe material constructed with new technology has high-performance properties compared to that constructed techniques that were used 80 years ago. Nevertheless, effective preservation techniques have appeared in the intervening years. Second, although the low performing properties of aged materials and their degradation during service (prior to cathodic maintenance, for example) is of particular concern, recent detections and examination of pipelines made of aged materials have been applied to find possible problems prior to the damage. Third, the successive acceptable operation of any pipeline, aged or recent, needs levels of detection and preservation suitable to the operating properties of the materials and the causative factors of degradation to which the pipeline is subjected in its functioning environment. Eventually, novel technology is able to recognize and detect small flaws, hence increasing efficiency further.

1.5 Evolution of Pipeline Technology

The first pipe systems in the United States were laid early in the nineteenth century to transfer manufactured gas for gas-lighting aims in big cities. The pipe systems were commonly made of cast-iron pipe that was a pipe made predominantly from gray cast iron manufactured in the United States in 1834. Pipelines manufactured of wrought iron and connected by screwed collars were as well utilized in gas usages. Soon after oil had been found in Pennsylvania in 1858, a completely new application for pipeline systems appeared. The earliest oil pipe system, a 2–1/2- miles long and in diameter of 2 in. worked successfully in 1863 such that was transferring 800 barrels of oil daily [30]. The threaded parts of the pipeline had been connected by screwed couplings. Techniques have improved in recent years and led to the reduction of many different types of risks in manufacturing and construction issues [29, 31]. These improvements are:

- Advances in the efficiency of the material and efficient means for pipe manufacture decreased the probability of damages in the pipeline material or the constructed longitudinal seam;
- Advances in the installation of pipe systems decreased the probability of damages in the connection points in the distance around the pipe;
- Advances in controlling the operators that generate flaw and damage in the service environment decreased the probability of damages because of the exterior corrosion;
- Advances in the examination, checking, and preserve pipes decreased the probability of damages because of a variety of reasons, even third-party flaws, the highest reason for pipeline safety accidents.

1.6 Evolution of Pipeline

For years, pipe systems have been made around the world to transfer water for drinking as well as irrigation purposes. This covers olden usage in China of pipeline constructed of hollow bamboo also the usage of flumes by the Romans as well as Persians. The Chinese furthermore applied bamboo pipes for transferring natural gas to their capital for lighting purposes. An important advance of piping device happened in the eighteenth century, while cast-iron pipes were widely manufactured and utilized. A further main milestone was the appearance of the steel pipeline in the nineteenth century that extremely increased the mechanical hardness of the piping material. The growth of steel pipelines with high hardness made it feasible to transfer natural gas as well as oil over vast distances. At first, the entire pipes made of steel needed to be threaded to fasten together. However, that was hard to perform on large pipelines, as they could leak under great pressure. In the 1920s, the welding machines were commonly utilized for making leakproof, high-pressure, large-diameter pipe systems. Nowadays, the majority of high-pressure pipeline systems are made of steel pipe with welded connections [1, 32].

Since 1950 most developments have been focused on the origination of the ductile iron as well as concrete pressure pipe systems with large diameter for water; application of polyvinyl chloride pipes for underground conduits; application of smart pig to remove dirt from the inside of pipes and to carry out other responsibilities; arranging various petroleum productions in a general pipe system; usage of cathodic preservation to decrease corrosion and increase the life of pipe system; application of space-age devices like computers for controlling the pipe systems and satellites for communicating between the enteral stations and the field; also modern devices and wide measures for preventing as well as identifying leakages in pipe systems. Moreover, many modern technologies have been developed or introduced to simplify pipeline structure for example application of new devices for drilling below rivers and roads to cross, application of new devices for bending extensive pipe systems in the field, and application of X-ray machines to identify welding damages [33].

1.6.1 Types of Pipeline

Pipelines have been classified in various ways based on the commodity transferred and also the kind of fluid flow.

1.6.1.1 Water and Sewer Lines

Pipeline systems have been utilized commonly to transfer water from treatment plants into the household or industry. Underground pipe networks are placed under the city streets and convey the water to almost every house. Water pipe systems are typically buried underground and the depth of cover is only a few feet, relying on the freezing depth of the position and the requirement for maintenance against unexpected injury by digging or manufacturing operations.

In recent water technology, when copper pipes have been typically utilized for indoor plumbing, outdoor water service lines with big diameter could utilize steel, ductile cast iron, or concrete pressure pipes. Outdoor water service lines with small diameter could utilize steel, ductile cast iron, or polyvinyl chloride pipes. In a case that metal pipelines are applied for transferring drinking water, the inside of the pipeline is usually made of a plastic or cement lining for preventing rusting, which could cause decay in water quality. The outside of metal pipelines is moreover covered with an asphalt material and twisted with a specific tape for decreasing corrosion because of the contact with specified soils. Furthermore, a surface of electrodes under a direct current are usually placed through steel pipes and is named cathodic preservation [34].

A sanitary sewer or foul sewer is a system of pipe and tunnel designed to transfer sewage from homes and buildings to treatment feasibilities or access. Foul sewers are considered as a part of the total system functions named as sewage system or sewerage. Sewage could be treated for controlling the impurity of water afore release to surface waters [35, 36]. Foul sewers serving commercial and industrial fields furthermore carry sanitary or industrial wastes. Separate sanitary sewer system design and technology has been only used for transporting sewage. However, in municipalities rendered services by sanitary sewer systems, discrete storm drains could transmit surface runoff straightly to an outlet. Sanitary sewer systems could be identified from merged sewers, that merge sewage with stormwater runoff in a single pipe. These systems are useful as they evade merged sewer overflows [37].

Domestic sewage is generally a mixture of 98% water and 2% solids. Transporting sewage by pipelines cause a slightly reduction on the internal diameter of the pipe, however it is below low pressure. Culvert or storm sewer with large diameter usually utilize corrugated steel pipe [38]. According to the pressure in the pipeline and further situations, sewer pipelines are composed of concrete, polyvinyl chloride, cast iron, or clay. Polyvinyl chloride is mainly desired for sizes smaller than 12 in. (30 cm) in diameter.

1.6.1.2 Oil Pipelines

There exist two kinds of oil pipe systems, namely: crude oil pipe system and product pipe system. Crude oil pipe system transfers crude oil to oil refining and refined product markets, and the product pipe system transfers refined products like gasoline, kerosene, jet fuel, and heating oil from oil refining and refined product markets to the market. Various types of crude oil or various refined products have been typically transferred via the same pipe system in various batches. Sometimes mixing is done in batches that are small and could be controlled. This can be done either by applying big batches or by keeping a ball among batches for splitting them. Crude oil and refined petroleum products transferring via pipe systems usually have a tiny quantity of preservatives to decrease interior corrosion of pipeline and diminish the loss of energy (drag- reduction preservative). The most widely known types of drag-reduction preservatives are polymers like polyethylene oxides. Oil pipe systems particularly utilize steel pipes without coatings. However, an exterior coating and cathodic preservation could be used to decrease exterior corrosion. They are welded together and bent to shape in the field. These pipes are joined together and curved to form in the field.

The Big Inch and Little Big Inch were two-pipe systems placed from Texas to New Jersey while World War II to combat the strike of German submarine tanker assaults on the coast. A long product pipe system placed from the Houston area to Linden, New Jersey, and constructed by the Colonial Pipeline Company in the 1960s to combat the threat of the maritime union. The Trans Alaska pipe system was constructed to carry crude oil from the North Slope to Prudhoe Bay to meet the 1973 oil crisis caused by the Arab oil embargo.

Subsea pipe systems are required to transfer oil as well as gas production from subsea oil and gas wells to onshore pipe systems that moreover transfer the oil to refineries and tanker loading facilities or the gas to a processing plant. Subsea pipe systems are more costly and complicated to construct compared with onshore pipe systems. Subsea manufacturing generally applies a barge on which pipe portions are successively joined together and attached to the onshore pipe endpoint. Because more portions are joined to the end of the pipeline, the barge goes to the oil or gas platforms, and the ending section of the pipeline is constantly moved down into the ocean at the back of the barge. Manufacturing advances up to the barge have arrived at the platform and the pipeline is attached to the oil or gas well. In deep oceans with rogue waves, ships in place of barges are employed for laying and setting of pipelines on the ocean floor. One of the most significant subsea oil pipe systems is located in the North Sea and connects the United Kingdom North Sea oil fields to the Shetland Islands.

1.6.1.3 Gas Pipelines

The gas pipeline transportation system is a system utilized to transfer gas from gas wells to the processing plants and subsequently to a gas distribution network. Natural gas is straightly transferred to homes or buildings via an extensive network of gas pipe systems. In practice, all onshore transport of natural gas has been done using a pipe system. Transportation of natural gas using other means like truck, train, or barge is extremely unsafe and costly. When gas gathering and transfer networks are constructed of steel, the majority of distribution systems, for example, small transmission line connection means joining the principle or transfer lines to consumers constructed in the United States since 1980 employ pliable plastic pipelines that could easily lay and also, they do not oxidize. The United States runs the greatest and most advanced natural gas pipeline distribution system in the world. Natural gas consumption has also increased considerably in other countries around the world.

1.6.1.4 Pipelines for Transporting Other Fluids

Pipe systems have been constructed for transporting several types of fluids. For example, liquid fertilizers can be traveled significant distances through pipe systems. The blend of oil and natural gas extracting from a well should be travelled as gas–liquid two-phase flows by pipe systems to onshore process plant before final oil/gas separation. Liquefied natural gas travelled by tanker ships with temperature-controlled tanks moreover needs small pipe systems to attach the tanker ships to supply tanks on land. There are 180-mile pipe systems in the United States to carry carbon dioxide to oil fields in West Texas to enhance production. Eventually, on a specific scope such as pharmaceutical or specialty chemicals production, pipelines have been used to carry different fluids and gases within the plants. As these liquids cannot stand impurities such as chlorides therefore, the pipeline should be made of inert materials [39].

1.6.1.5 Slurry Pipelines

A slurry pipeline is a pipe system that is engineered to carry ores, like coal or gold, or copper, over vast distances. The slurry is a mix of the ore particles and water and is pumped to its destination. At its destination, the coal is removed from the water. Because of the corrosive features of slurry, the pipe systems could be constructed totally from high-density polyethylene pipes [40]. However, this may need an extremely thick-walled pipe. Slurry pipe systems have been employed as a replacement for railroad transport in a case that mines are placed in far, out of reach regions [41].

Investigators at the University of Alberta, located in Edmonton, Alberta, Canada are studying the application of slurry pipe systems to take agricultural and forestry residues from many dispersed resources to a central biofuel plant. For distances greater than 100 km pipelines option was found to be more viable to transport biomass (i.e., charcoal, pellets). In comparison with an identically sized oil pipe system, a biomass slurry pipe can transport nearly 8% of the product [42].

The slurry is a mix of the solid particles of iron ore and a liquid, typically water. Typically, particles lie within the range in size from larger than 4 in. in identical diameter to under 1/1000 of an inch. If the solid particles of iron ore in the water are tiny the blend is usually named fine slurry, also if the solid particles of iron ore in the water are big the blend is usually named coarse slurry. The conventional mining companies have used pipe systems to carry mine wastes to the waste disposal site in the form of a slurry, employing water as the liquid. Sometimes sand, clay, or silt is dredged from the seabed with water via a pipe system to a manufacturing site that is some distance from.

Generally, in a case that pipe systems are employed to carry coarse slurry, the slurry speed should be comparatively great for delaying the solids. This kind of slurry transportation is extremely corrosive to the pipeline, also the energy consumption is roughly at a high level. Accordingly, coarse slurry pipe systems are cost-effective on short journeys and are not reasonable for long distances. Ultimately, at the endpoint of the pipe, an industrial filter press is used to separate solid particles in a slurry from water. Generally, this water needs to undergo a wastewater treatment process before that goes back to the mine site. There is a considerable cost advantage of slurry pipe systems over railroad transportation. Furthermore, slurry pipe systems provide high noise reductions compared to railroad transportation, especially in a case that mines are located in far regions.

Pipe networks should be efficiently designed to have good abrasion resistance as well as high corrosion stability. High-density polyethylene pipe lining systems can be used for internal pipeline protection. Slurry transfer pipe systems are normally used to carry coal, copper concentrate, iron concentrates, limestone, lead, kaolin, salt in the brine, and oil sands [43]. Slurry pipe systems have been moreover employed to carry waste from a mineral processing plant to a tailings storage facility after ore extraction. A mixture of oil sand and process water from ore preparation plants could be pumped through a long-distance hydrotransport pipeline system. Usually, long-distance transportation of solids using a slurry pipe system would need to employ relatively fine slurry. Current Hypothetical coal slurry pipe systems transport fine slurry composed of approximately 50% coal as well as 50% water by weight. Initially, the solid forms a paste when it is crushed and blended with water. The slurry afterward goes into a blending tank, that is made of one or more big rotating wheels to make the particles uniformly blended. As a next step, the slurry goes into the pipe system. The specific type of pumps (e.g., plunger or piston pumps) can be utilized to transport the slurry over long distances.

1.6.1.6 Pneumatic Pipelines

The pneumatic tube system is technical equipment in which cylindrical containers are propelled through the system tubes using pressurized air or partial vacuum [44]. These systems have been utilized to transport solid materials, as opposed to ordinary pipe systems that carry fluids. Pneumatic tube systems achieved acceptance around 1900 for offices that wanted to send medicines, cash packets, and other small items over very small distances, such as within a building or the inner-city area. Some installations grew to high intricacy, however, they were mainly replaced. More recently, the usage of pneumatic tube systems in some settings, such as hospitals have been remarkably extended [45].

There exist two common kinds of pneumatic pipe systems [46]. The initial kind uses suction lines to produce suction in the pipeline by keeping the air compressor close to the downstream termination of the pipeline. The latter kind is pressure lines, and have compressors placed close to the upstream termination of the pipeline. This generates a pressure in the line and makes the air as well as the solids to be traveled overall the pipeline. Pressure transmission lines have been employed for greater distances and in cases that solids concentrated at one place are sent to various distinct places applying an individual blower or compressor. Conversely, suction lines have been employed for smaller distances and in cases that solids from various places are sent to a general destination using an ordinary blower or compressor [47]. A pneumatic pipe network should further contain a tank or hopper joint near the pipe entry to supply solid particles into the pipe system and a tank close to the pipe exit to split the conveyed solids from the airstream. The outlet air moreover should be filtered to stop air impurity.

Systems of pneumatic pipelines could be used to carry combustible solids such as grain, flour, coal, or gunpowder, however, if they handle incorrectly, could lead to fire or even explosion. This is due to the accumulation of electric charges on fine particles transported pneumatically. The electric charges accumulated on solid particles can cause such fire or explosion. To prevent such fire or explosion some schemes can be used such as using metal pipelines instead of plastic pipelines, installing underground pipelines, cleaning the inside of the pipeline to remove the dust, and enhancing the humidity of the air employed for pneumatic transportation.

1.6.1.7 Capsule Pipelines

Capsule pipelines technology can transport freight such as coal and solid waste which uses air to push the capsules through the pipeline. The capsule pipe systems can carry large amounts of freight for great distances in both horizontal and vertical directions having lower operating costs than its competitors. The wheeled capsules are regularly filled with ore, concentrate, coal, solid waste, and directed via a pipe system to be automatically stopped to drop off their contents at the receiving terminal [48]. Wheeled capsules are attached in trains and run within the pipe system concurrently therefore the transportation system is continuous. Typically, wheeled capsules can transfer multiple products without mixing and with a very low operating cost.

Capsule pipe systems carry freight in capsules directed by a fluid proceeding via a pipe system. If the conveying fluid is a gas (namely air), the technology can be named pneumatic capsule pipeline [49], and, if the conveying fluid is water, the technology can be named hydraulic capsule pipeline [49]. Due to the low air density, capsules in the pneumatic capsule pipeline may not be stopped by air at normal velocities. Alternatively, capsules (wheeled vehicles) roll throughout the pipe systems. Conversely, as water is heavy, the capsules in the hydraulic capsule pipeline do not need wheels. The pneumatic capsule pipeline and hydraulic capsule pipeline are both moved and stopped by water under normal operating velocities. Hydraulic capsule pipeline systems typically function at a velocity of 6–10 feet per second, whilst the functional velocity of pneumatic capsule pipeline is typically much greater—20 to 50 feet per second. Due to high friction losses at high speeds, pneumatic capsule pipelines use more energy in operation in comparison with hydraulic capsule pipelines [50].

Pneumatic capsule pipelines have been employed since the nineteenth century to transport a wide variety of products such as mail, printed telegraph messages, machine segments, and samples of blood (in hospitals). Since 1970, great wheeled pneumatic capsule pipelines have been produced to transport weighty cargo over comparatively great distances. The greatest pneumatic capsule pipeline in the world is LILO-2 placed in the republic of Georgia about 11 miles long and 48 in. diameter and is mainly constructed to transport rock [1, 51].

There's a striking contrast between pneumatic capsule pipeline systems which are widely used and hydraulic capsule pipeline systems which is in its early stage of growth or development. Hydraulic capsule pipeline systems were initially employed by the British army to transport military equipment in East Asia while the Second World War. Extensive research was carried out on capsule pipeline technology in Canada at the Alberta Research Council from 1958 to 1975. The spread of this new technology within countries was significantly high. In 1991, the National Science Foundation in the United States created a Capsule Pipeline Research Center at the University of Missouri-Columbia [52]. A new structure of the hydraulic capsule pipeline system is the coal-log pipeline [53], which carries compacted coal logs. The coal-log pipeline removes the usage of capsules for enclosing coal and the requirement to have a distinct pipe system for returning vacant capsules. For the same diameter, the coal-log pipe system carries more coal utilizing less water in comparison with the coal-slurry pipe system [1, 54].

Large-diameter capsule pipe systems have been successfully employed to carry most types of cargo typically transported by trucks or trains. In both Europe and the United States, capsule pipe systems of large diameter (mainly pneumatic capsule pipelines) are suggested for freight transportation services [51, 55]. Such capsule pipe systems for freight transportation services not just permit the face of the earth to be utilized for other aims but they as well decrease the number of trucks as well as trains required, which for their part decrease air impurity, accidents, and harm to highway and rail foundations due to the heavy traffic [56].

1.7 Design and Operation

Many factors affect the pipeline design such as the choice of the path traveled by the pipeline, specification of the construction and the functioning speed, computing of pressure gradient, the choice of pumps and other instruments, and determining the pipe sizes and equipment. The design experts should give serious consideration to security, preventing leak, blockage and damage, government laws and regulations, and environmental challenges and impacts.

1.7.1 Components

The pipe systems are made of various pieces of materiel that function simultaneously to carry products from one place to another place. The major components of a pipe network are primary injection building, compressor and pumping buildings, partial delivery or intermediate building, block valve building, regulator building, terminal delivery building. Particular pipe systems that carry cryogenic liquids, for example liquid nitrogen, liquid helium, and liquid carbon dioxide, should have refrigeration systems to maintain the liquid in the pipeline under the lower critical temperature [57, 58].

1.7.2 Construction

The preparation and design of pipe systems require a detailed survey of the route, ditching, and excavation, hauling the pipes, and other equipment to the location, placing pipes in assembly position alongside the ditch centerline, curving steel pipes in the field to fit a required alignment using the pipe-bending machine, coating of steel pipes and fittings for corrosion protection, welding pipes together before or after being placed in the trench, ensuring the satisfactory performance of a welded structure, and then refilling the trench by soil and restoration to return the land to its original condition. Long-distance pipelines have to be made in segments therefore, construction of the second segment starts once the construction of the first segment is completed and so on. Hence, the time that any given place is allocated during construction activities would be minimized. Even when the pipe systems are great, building for any segment is normally ended within six months and mostly in a short time. Small pipe systems could be built in days. Typically, while passing a pipeline through a river or stream, the pipeline can be either connected to a bridge structure, placed along the stream bed or bored through the soil below the stream. A tunnel boring machine can be used when pipelines need to cross rivers and roads.

1.7.3 Operation

Large modern pipe systems are designed to operate automatically by a centralized computer network control system at the pipeline company. The centralized computer network control system observes the rate of flow, pressure, discharge of liquids or gasses and other parameters at different places along the pipeline, carries out numerous online calculations and sends corrective signals to the field devices to control the function of the valves and compressors. Human intervention is often required to changing operating modes, whenever various batches of fuels are sent to various containers designed for the temporary storage of fuels, or whenever the system should be turn off or turn on.

1.7.4 Safety

The safety and reliability of pipe networks rely heavily on the materials transported. Recently, for making proper decisions about the safety and integrity of the pipes, the risk values and risk factors are mainly important issues to be discussed. Nevertheless, the challenge is the validity of the models used to obtain the risk data. Hence, there is a requirement for a potent device to deal well with that uncertainty. Possibly the best device for coping with uncertainty is the application of artificial intelligence techniques utilizing fuzzy logic. Artificial intelligence technique has proven to be highly effective in many fields [59–80] especially in manufacturing systems. Pipeline operators employ several techniques for pipeline safety and also to detect faults in the pipelines [81–85]. Pipe networks that carry water or employ water to carry coarse-grain solids, like hydraulic capsule pipe systems, do not cause any environmental pollution in case that the pipe breaks or tears. The rupture of crude oil (petroleum) pipes does not result in a fire, however, they can deposit the pollution into water and soil. Natural gas pipes and product pipes that carry highly volatile liquids, such as butane, ethane, propane can be easily exploded because of a pipeline leak, therefore they deserve special consideration and regulatory treatment. Despite the fact that there are many different risk factors or threats for pipelines, they are proven to be the safest way to transport petroleum and natural gas products. Apparently using other modes of transportation such as truck or railroad to move such fuel on land are very risky and expensive.

During the decade after the Second World War, numerous large pipe networks were built. The boom in pipe manufacturing led to numerous inventions and technological progress in pipeline construction and building. Longitudinal submerged arc welded pipes manufacturing process was one the most common approach to construct large diameter pipes [86]. Later double submerged arc welding was used that mechanically expanded the pipe. The double submerged arc welding showed to be more valid and was widely accepted as the unique tools of constructing submerged arc welded pipes. Immediately after welding, the cold expansion was applied to every piece of submerged arc welded pipe to make the diameter of the pipe uniform [30].

Although pipelines have had a good record than other transportation modes, in the United States there is a great concern about pipe safety as spills and incidents still happening. In the United States the main emphasis has been placed on pipeline safety. Much research has been carried out to prevent pipeline rupture or leak accidents and also to rectify problems anytime they happen. Analysis of pipeline accidents in the United States has been shown that the third-party damage has become the main cause of nearly half of all pipeline incidents, as, for example, a builder could damage the pipe when digging the home's foundation. As a result, pipeline companies have a specific education program for the public on pipeline security and share information to building and infrastructure groups on the places of buried pipelines for decreasing third-party damage.

Pipeline corrosion is the number one factor in pipeline failure, which is an electrochemical process resulting from an electrochemical reaction between metal surfaces with wet soil. Pipeline security and other actions taken by pipeline companies to combat corrosion are coating buried pipes with tape and employing cathodic protection to protect outer corrosion also inserting specific chemicals to the fluid against inner corrosion. Hydrazine and sodium sulfite are the two most widely used chemicals for controlling corrosion in water pipelines. The decrease in corrosion is due to the reaction of these chemicals and therefore eliminating most of the dissolved oxygen from water.

Eventually, the leakage detection can be carried out by computer observation of unusual flow rates and pressure also using light aircraft or helicopters along the pipeline route for visual detection. Furthermore, Smart pigs can be sent through pipelines to identify corrosion, weak welds, and other signs of damage.

1.8 Pipeline Milestones

1834	Initial North American cast iron pipeline made at Millville, New Jersey
1856	Development of the Bessemer steel
1858	Initial American oil well made at Titusville, Pennsylvania
1863	Initial oil pipe network transported 800 tanks of crude oil daily
1863	Screwed couplings used to join pipes and fittings
1863	Development of wrought iron piping
1869	Development of pipeline hydrostatic testing to maintain the quality and safety of the pipe
1871	Displacement of wrought iron by the adoption of Bessemer steel
1891	Development of Dresser coupling for joining two pieces of pipes
1897	Construction of the initial 30 diameters lap-welded pipeline
1899	Construction of the initial 20 diameters seamless pipeline
1900	Maximum of lap-welded pipelines were formed from steel sheets
1904	The initial large-diameter gas transportation pipe network
1911	Acetylene girth welds were used to build a 1-mile pipe network
1914	Acetylene girth welds were used to build a 35-mile pipe network
1917	Electric metal arc welding was used to build an 11-mile pipe network
1919	American Petroleum Institute, the major trade association of the oil industry established
1924	Low-frequency electric resistance welding current was used to build pipeline
1925	Development of large-diameter seamless pipeline
1927	Electric flash welded pipeline was manufactured
1928	Development of American Petroleum Institute Standard 5L (Electric Fusion Welding, electric resistance welding, spiral submerged arc welding, longitudinal submerged arc welded) for manufacturing pipeline
1930	Electric arc girth welding was used to build the first long-distance pipe network
1931	An extension occurred, upon American Petroleum Institute Standard 5L that was including electric resistance-welded pipe
1933	Maximum of pipe networks were welded with electric arc girth welding
1935	Publishing of initial American tentative pressure piping code
1942	An extension occurred, upon American Petroleum Institute Standard 5L that was including hydrostatic testing of pipeline
1942	Publishing of American standard pressure piping code
1942–1943	Construction of crude oil line and products line from Texas to New Jersey during the Second World War
1944	An extension occurred, upon American Petroleum Institute Standard 5L that was including electric flash-welded pipeline
1946	Construction of single submerged arc welding pipeline

(continued)

(continued)

1834	Initial North American cast iron pipeline made at Millville, New Jersey
1948	Introduction of radiographic testing for the inspection of girth welds
1948	Construction of double submerged arc welding pipeline
1948	Development of American Petroleum Institute Tentative Standard 5LX for manufacturing pipeline
1951	Pressure piping code was approved by the American National Standards Institute
1953	American piping products in grades X46 and X52 were presented
1956	Hydrostatic mill test was presented
1959	First distinct code for oil pipeline transportation
1962	Manufacturing of furnace lap welded pipeline was discontinued and process deleted from American Petroleum Institute 5L
1962	Oxygen converter process accepted in American Petroleum Institute 5L
1963	Pipe body nondestructive examinations started to use in American Petroleum Institute 5L pipe specification
1965	Initial application of intelligent pig in the pipe system
1966	Development of American Petroleum Institute 5L grade X60 pipeline
1967	Development of American Petroleum Institute 5L grade X65 pipeline
1969	Manufacturing of furnace lap-welded pipeline discontinued and process deleted from American Petroleum Institute 5L
1969	Supplementary fracture toughness tests added in American Petroleum Institute 5L
1970	Manufacturing of Bessemer steel discontinued and process deleted from American Petroleum Institute 5L
1970	Federal liquid pipeline safety standards were published
1973	Development of American Petroleum Institute 5L grade X70 pipeline
1977	Trans-Alaska pipeline system began operation
1980	Application of high-resolution intelligent pig in the pipe system
1982	Intelligent pig inspection tested
1983	American Petroleum Institute 5L and American Petroleum Institute 5LX merged in American Petroleum Institute 5L
1985	Development of American Petroleum Institute 5L grade X80 pipeline
1992–95	Testing crack inspection devices in a variety of applications
2000	The lowest fracture toughness value was proposed in American Petroleum Institute 5L
2001	Initiating pipeline integrity management system design for dangerous liquid carrying pipe systems

References

1. Needham, J.: Science and Civilization in China, vol. 4, p. 33. Caves Books Ltd., Taipei (1986)
2. The Forrest family Archived 2016-08-17 at the Wayback Machine Dynasties (2006). ABC
3. Mannum Adelaide Celebrations SA Water. Archived from the original on 2015-05-03. 2015 IEEE Student Symposium in Biomedical Engineering and Science (ISSBES)
4. GMR (Great Man-Made River) Water Supply Project, Libya (2012). Water-technologynet Retrieved Apr. 15
5. Morgan-Whyalla Pipeline Bill.: The Advertiser Adelaide: National Library of Australia (23 Aug. 1940), 20 (2014)
6. Hammerton, M.: Water South Australia: A History of the Engineering and Water Supply Department. Wakefield Press (1986)
7. The World Factbook—Central Intelligence Agency (2016). Archived from the original on August 21, 2016 Retrieved September 6
8. Pipeline transport (2015). Retrieved 26 January 2015
9. The Transportation of Natural Gas (2019). Retrieved 2019-07-18
10. Waldman, J.: How the Oil Pipeline Began. Nautilus (Science Magazine) (2017)
11. Conca, J.: Pick Your Poison For Crude -- Pipeline, Rail, Truck Or Boat. Forbes
12. American Petroleum Institute. Accessed 20 Feb. 2010
13. Drilling Productivity Report.: US Energy Information Administration (2017)
14. Line 9: Journey along the pipeline|Toronto Star (2015)
15. Ulvestad, M., Overland, I.: Natural gas and CO_2 price variation: impact on the relative cost-efficiency of LNG and pipelines. Int. J. Environ. Stud. **69**(3), 407–426 (2012). https://doi.org/10.1080/00207233.2012.677581
16. Enbridge Sandpiper Pipeline (2015)
17. Oil & Gas Pipeline Construction in the U.S.: Market Research Report. IBIS World (2012)
18. 2012 Worldwide Pipeline Construction Report Archived 2013-03-25 at the Wayback Machine. Pipeline Gas J. **239**(1) (2012)
19. Mohitpour, M.: Pipeline Design and Construction: A Practical Approach. ASME Press (2003)
20. go-devil—definition of go-devil by the Free Online Dictionary. Thesaurus and Encyclopedia
21. Kiefner, J.F., Kiefner, B.A., Vieth, P.H.: Analysis of DOT Reportable Incidents for Hazardous Liquid Pipelines, 1986 Through 1996. API Publication 1158 (1999)
22. Trench, C.J.: The U.S. Oil Pipeline Industry's Safety Performance (2001)
23. Laird, C.: The influence of metallurgical structure on the mechanisms of fatigue crack propagation. In: Grosskreutz J (ed) Fatigue Crack Propagation. ASTM International, West Conshohocken, PA, pp. 131–180 (1967). https://doi.org/10.1520/STP47230S
24. Dreyfuss, G., Smith, A.A.: Automatic welding of pipelines with the 'saturne' process on a Laybarge. In: Welding in Energy-Related Projects. Pergamon, pp. 115–122 (1984). https://doi.org/10.1016/B978-0-08-025412-8.50016-4
25. Huntley, R.M., Dorling, D.V., Rothwell, A.B.: Pipeline girth welding using the flux-cored arc welding process. In: Welding in Energy-Related Projects. Pergamon, pp 85–94 (1984). https://doi.org/10.1016/B978-0-08-025412-8.50013-9
26. Fink, J.: Chapter 7 - Pipeline Cleaning. In: Fink J (ed) Guide to the Practical Use of Chemicals in Refineries and Pipelines. Gulf Professional Publishing, Boston, pp. 109–129 (2016). https://doi.org/10.1016/B978-0-12-805412-3.00007-6
27. Sasseen, K.M., Chilingarian, G.V., Robertson, J.O.: Chapter 1 Introduction to Surface Production Equipment. In: Chilingarian, G.V., Robertson, J.O., Kumar, S. (eds) Developments in Petroleum Science, vol. 19. Elsevier, pp. 1–41 (1987). https://doi.org/10.1016/S0376-7361(08)70530-2
28. Stewart, M.: Design of Gas-Handling Systems and Facilities. Surface Production Operations, vol. 2, 3rd edn. Elsevier, Waltham, Massachusetts (2014)
29. Caretta, M.A., McHenry, K.A.: Pipelining Appalachia: a perspective on the everyday lived experiences of rural communities at the frontline of energy distribution networks development. Energy Res. Soc. Sci. **63**, 101403 (2020). https://doi.org/10.1016/j.erss.2019.101403

30. Milward, A., Saul, S.B.: The Development of the Economies of Continental Europe 1850–1914., pp. 1–96 (2012)
31. Scott, R.P., Scott, T.A.: Investing in collaboration for safety: assessing grants to states for oil and gas distribution pipeline safety program enhancement. Energy Policy **124**, 332–345 (2019). https://doi.org/10.1016/j.enpol.2018.10.007
32. Lawal, M.: Historical development of the pipeline as a mode of transportation. Geogr. Bull. **43**(2), 91–99 (2001)
33. Twentyman, M., Rosetti, R., Porta, G.: Microstructural evolution of pipelines for thermal electric power plants after a prolongated operation
34. Özer, A., Kasirga, E.: Substrate removal in long sewer lines. Water Sci. Technol. **31**(7), 213–218 (1995)
35. Whipple, G.C.: Sewerage and Sewage Disposal, a TextBook. American Public Health Association (1922)
36. Allen, K.: Sewerage and Sewage Disposal. American Public Health Association (1930)
37. Selvakumar, A., Field, R., Burgess, E., Amick, R.: Exfiltration in sanitary sewer systems in the US. Urban Water J. **1**(3), 227–234 (2004)
38. Swamee, P.K.: Design of sewer line. J. Environ. Eng. **127**(9), 776–781 (2001)
39. Mokhatab, S., Mak, J.Y., Valappil, J.V., Wood, D.A.: Handbook of Liquefied Natural Gas. Gulf Professional Publishing (2013)
40. Adhikary, K.B., Pang, S., Staiger, M.P.: Dimensional stability and mechanical behaviour of wood–plastic composites based on recycled and virgin high-density polyethylene (HDPE). Compos. B Eng. **39**(5), 807–815 (2008)
41. Al-Salem, S., Lettieri, P.: Kinetic study of high density polyethylene (HDPE) pyrolysis. Chem. Eng. Res. Des. **88**(12), 1599–1606 (2010)
42. Postlethwaite, J., Tinker, E., Hawrylak, M.: Erosion-corrosion in slurry pipelines. Corrosion **30**(8), 285–290 (1974)
43. Cerisier, P., Porterie, B., Kaiss, A., Cordonnier, J.: Transport and sedimentation of solid particles in Bénard hexagonal cells. Eur. Phys. J. E **18**(1), 85–93 (2005)
44. Van, O.W.: Pneumatic tube system. Google Patents (1973)
45. Lang, H.: Pneumatic tube conveyor system. Google Patents (1993)
46. Beltrop, H., Teutenberg, J., Hilbig, M.: Pneumatic tube installation for posting samples of material. Google Patents (1983)
47. Grosswiller, L., Anders, W.G., Mannella, L.F.: Pneumatic tube system. Google Patents (1994)
48. Kruyer J, Redberger P, Ellis H (1967) The pipeline flow of capsules. Part 9. Journal of Fluid Mechanics 30 (3):513–531
49. Tomita, Y., Yamamoto, M., Funatsu, K.: Motion of a single capsule in a hydraulic pipeline. J. Fluid Mech. **171**, 495–508 (1986)
50. Feng, J., Huang, P., Joseph, D.: Dynamic simulation of the motion of capsules in pipelines. J. Fluid Mech. **286**, 201–227 (1995)
51. Liu, H.: Pneumatic capsule pipeline basic concept, practical considerations, and current research. In: Mid-Continent Transportation Symposium 2000 Proceedings. Citeseer (2000)
52. Liu, H., Zuniga, R., Richards, J.L.: Economic Analysis of Coal Log Pipeline Transportation of Coal. Capsule Pipeline Research Center (1993)
53. Liu, H., Assadollahbaik, M.: Feasibility of using hydraulic capsule pipeline to transport coal. J. Pipelines **1**(4), 295–306 (1981)
54. Liu, H., Marrero, T.R.: Coal log pipeline technology: an overview. Powder Technol. **94**(3), 217–222 (1997)
55. Liu, H.: Feasibility of using pneumatic capsule pipelines in New York City for underground freight transport. In: Pipeline Engineering and Construction: What's on the Horizon?, pp. 1–12 (2004)
56. Liu, H., Marrero, T.: Coal Log Pipeline: A New Process to Transport and Burn Coal. Missouri Univ, Columbia (USA) (1988)
57. Gorton, I., Wynne, A., Liu, Y., Yin, J.: Components in the Pipeline. IEEE Softw. **28**(3), 34–40 (2011)

58. Potter, S.C., Clarke, L., Curwen, V., Keenan, S., Mongin, E., Searle, S.M., Stabenau, A., Storey, R., Clamp, M.: The ensemble analysis pipeline. Genome Res. **14**(5), 934–941 (2004)
59. Yu, W., Jafari, R.: Modeling and Control of Uncertain Nonlinear Systems with Fuzzy Equations and Z-Number. Wiley (2019)
60. Razvarz, S., Jafari, R.: ICA and ANN modeling for photocatalytic removal of pollution in wastewater. Math. Comput. Appl. **22**(3), 38 (2017a)
61. Jafari, R., Razvarz, S., Gegov, A.: Neural network approach to solving fuzzy nonlinear equations using Z-numbers. IEEE Trans. Fuzzy Syst. (2019)
62. Razvarz, S., Jafari, R.: Intelligent techniques for photocatalytic removal of pollution in wastewater. J. Electr. Eng. **5**(1), 321–328 (2017b)
63. Jafari, R., Razvarz, S., Gegov, A., Paul, S., Keshtkar, S.: Fuzzy Sumudu transform approach to solving fuzzy differential equations with Z-numbers. In: Advanced Fuzzy Logic Approaches in Engineering Science. IGI Global, pp. 18–48 (2019)
64. Jafari, R., Razvarz, S., Gegov, A.: A novel technique to solve fully fuzzy nonlinear matrix equations. In: International Conference on Theory and Applications of Fuzzy Systems and Soft Computing, 2018, pp. 886–892. Springer (2018)
65. Jafari, R., Razvarz, S., Gegov.: A fuzzy differential equations for modeling and control of fuzzy systems. In: International Conference on Theory and Applications of Fuzzy Systems and Soft Computing, 2018, pp. 732–740. Springer (2018)
66. Jafari, R., Yu, W., Razvarz, S., Gegov, A.: Numerical methods for solving fuzzy equations: a survey. Fuzzy Sets and Systems (2019)
67. Jafari, R., Razvarz, S., Gegov, A.: A new computational method for solving fully fuzzy nonlinear systems. In: International Conference on Computational Collective Intelligence, 2018, pp. 503–512. Springer (2018)
68. Jafari, R., Razvarz, S., Gegov, A., Paul, S.: Modeling and control of uncertain nonlinear systems. In: 2018 International Conference on Intelligent Systems (IS), 2018, pp. 168–173. IEEE (2018)
69. Jafari, R., Razvarz, S., Gegov, A.: A novel technique for solving fully fuzzy nonlinear systems based on neural networks. Vietn. J. Comput. Sci. **7**(1), 93–107 (2020)
70. Razvarz, S., Hernández-Rodríguez, F., Jafari, R., Gegov, A.: Foundation of Z-Numbers and Engineering Applications. In: Latin American Symposium on Industrial and Robotic Systems, 2019, pp. 15–24. Springer (2019)
71. Jafari, R., Contreras, M.A., Yu, W., Gegov, A.: Applications of fuzzy logic, artificial neural network and neuro-fuzzy in industrial engineering. In: Latin American Symposium on Industrial and Robotic Systems, pp. 9–14. Springer (2019)
72. Jafari, R., Razvarz, S., Gegov, A., Yu, W.: Fuzzy control of uncertain nonlinear systems with numerical techniques: a survey. In: UK Workshop on Computational Intelligence, 2019, pp. 3–14. Springer (2019)
73. Jafari, R., Razvarz, S., Yu, W., Gegov, A., Goodwin, M., Adda, M.: Genetic algorithm modeling for photocatalytic elimination of impurity in wastewater. In: Proceedings of SAI Intelligent Systems Conference, 2019, pp. 228–236. Springer (2019)
74. Tatchum, M., Gegov, A., Jafari, R., Razvarz, S.: Parallel distributed compensation for voltage controlled active magnetic bearing system using integral fuzzy model. In: 2018 International Conference on Intelligent Systems (IS), 2018, pp. 190–198. IEEE (2018)
75. Razvarz, S., Jafari, R., Gegov, A.: Solving partial differential equations with Bernstein neural networks. In: UK Workshop on Computational Intelligence, 2018, pp. 57–70. Springer (2018)
76. Jafarian, A., Jafari, R.: New iterative approach for solving fully fuzzy polynomials. Int. J. Fuzzy Mathe. Syst. **3**(2), 75–83
77. Jafarian, A., Jafari, R.: New method for solving fuzzy polynomials. Adv. Fuzzy Mathe. **8**(1), 25–33 (2013)
78. Jafarian, A., Jafari, R.: An iterative method for solving fuzzy polynomials by fuzzy neural networks (2012)
79. Jafarian, A., Jafari, R.: Simulation and evaluation of fuzzy polynomials by feed-back neural networks (2012)

80. Jafari, R., Yu, W.: Fuzzy control for uncertainty nonlinear systems with dual fuzzy equations. J. Intell. Fuzzy Syst. **29**(3), 1229–1240 (2015)
81. Razvarz, S., Vargas-Jarillo, C., Jafari, R., Gegov, A.: Flow control of fluid in pipelines using PID controller. IEEE Access **7**, 25673–25680 (2019)
82. Razvarz, S., Vargas-Jarillo, C., Jafari, R.: Pipeline monitoring architecture based on observability and controllability analysis. In: 2019 IEEE International Conference on Mechatronics (ICM), 18–20 March 2019, pp. 420–423 (2019)
83. Razvarz, S., Jafari, R., Vargas-Jarillo, C., Gegov, A., Forooshani, M.: Leakage detection in pipeline based on second order extended Kalman filter observer. IFAC-PapersOnLine **52**(29), 116–121 (2019). https://doi.org/10.1016/j.ifacol.2019.12.631
84. Razvarz, S., Jafari, R., Vargas-Jarillo, C.: Modelling and Analysis of Flow Rate and Pressure Head in Pipelines. In: 2019 16th International Conference on Electrical Engineering, Computing Science and Automatic Control (CCE), 11–13 Sept. 2019. pp. 1–6 (2019)
85. Jafari, R., Razvarz, S., Vargas-Jarillo, C., Yu, W.: Control of flow rate in pipeline using PID controller. In: 2019 IEEE 16th International Conference on Networking, Sensing and Control (ICNSC), 9–11 May 2019, pp. 293–298 (2019)
86. Duncan, I.J., Wang, H.: Improvements in pipeline failures after World War II: Reply. Int. J. Greenhouse Gas Control **42**, 700 (2015). https://doi.org/10.1016/j.ijggc.2015.09.008

Chapter 2
A Review on Different Pipeline Defect Detection Techniques

2.1 Introduction

Piping systems are the safest and most efficient and cost-effective way for fluid transportation around the world. Pipelines act as the most important type of transport infrastructure to keep our country moving. There exist 2.4 million miles of pipeline for transportation around the world. Pipe networks have been normally employed to transport crude oil from the production regions to distribution centres. Pipe systems are not only safer but need less energy to function than alternative transportation options. Nevertheless, apart from the industrial production processes, pipelines have been further used in aircraft hydraulic and fuel systems. Normally, transmission pipelines function up to 5000 psi [1], and up to 1400 PSIG for natural gas pipe systems [2]. As a result, circular pipe sections are often utilized because of the stability, strength, and rigidity of the structure and the uniform shape of the cross-section. Pipe networks are beneficial to transfer water for drinking purpose or irrigation at a great distance whenever it requires to go up and down hills, or where canals or channels are wrong options because of the considerations of vaporization, pollution, or environmental effect. Oil leaks in pipelines could cause a lot of damage to the environment and accordingly lead to explosions, fires, or injuries of the pipeline network. Direct impacts resulting from pipeline damage are production loss, loss of life, injury, or other health impacts. Moreover, blockage of pipeline network systems will lead to pressure build-up along the line and eventual rupture and explosion in case that it is not properly checked and fixed.

Leakages and blockages in pipe systems can be caused by several potential causal factors like using imperfect and substandard materials in the construction of pipelines, extreme functioning temperature of the pipeline, and fluid pollution caused by excessive biofilm build-up. Other types of pipeline damage include operations and production outside design basis, corrosion and wear, indeliberate third-party injury, and deliberate injury [3]. Corrosion is recognized as a main cause of failure in onshore gas, crude oil and refined fuels transferring pipe systems in the United States and was responsible for almost 18 percent of pipe accidents (both

S. Razvarz et al., *Flow Modelling and Control in Pipeline Systems*, Studies in Systems, Decision and Control 321, https://doi.org/10.1007/978-3-030-59246-2_2

onshore and offshore) between 1988 and 2008 [4]. Hence, pipe networks should be constructed with leakage and blockage inspection devices so that operators can be notified when the systems require detection [5]. To minimise the occurrence of pipe failure, leakage and blockage must be identified in their early stages therefore, necessary actions could be identified and implemented in a timely manner.

2.2 Non-destructive Testing Techniques for Flaw Identification in Pipelines

There exist several different non-destructive testing techniques recently utilized for the inspection of leakage and blockage faults in pipe networks and without having to interfere with the operating of the pipelines. However, those techniques have their limitations and weaknesses. Some recent non-destructive testing techniques employed to identify the fault in the pipeline are [6] ultrasonic, magnetic particle inspection, pressure transients, and acoustic wave approaches. Some approaches are positioning methods, and focus on sensors or dye scanning over each section of the surface of the pipeline during the detection process. However, it can be a daunting task with a time-consuming process. Moreover, some of those techniques are not always capable to detect leaks effectively and may fall in predicting the size and position of a defect [7]. Besides, radiography and safety equipment and facilities for the detection process are highly costly. It is important to note that apart from the equipment expense, radiography is a high-cost non-destructive testing technique, both in capital items, consumables, and manpower [8].

The pressure transients [9–17] and acoustic wave approaches [18–22] are remarkably inexpensive and effective techniques that can observe the impact of a flaw in a pipe network via observing the pipe reply. Accordingly, the acoustic wave propagation technique in a joint time–frequency domain has been employed to identify leaks and blockages in a pipe network filled with fluid. The acoustic wave reflectometry is a time domain-based approach and the roving mass is a frequency domain-based approach. The roving mass technique has been effectively utilized for crack, leakage, and blockage inspection in a pipe network filled with fluid.

2.3 Acoustic Wave Reflectometry and Roving-Mass Technique

The acoustic wave reflectometry technique involves the introduction of a pulse of pressure into the pipeline filled with fluid. The impacted pressure pulse causes the pipe system to shake, sometimes violently, and consequently generates wave propagation through the pipe such that the speed of propagation depends basically on the medium's characteristics. Nevertheless, if there exist any discontinuities in the cross-sectional area of the pipeline, from any cause like leakage, obstruction, and elbows, it will give rise to the occurrence of partial reflection and transmission of the sound wave on the joining points. At the same time, microphones that are installed throughout the pipeline have been employed to evaluate the transmission and reflection acoustic waves as they travel through the pipelines.

The roving mass technique is based on traveling the roving mass throughout the pipeline filled with fluid under acoustic excitation. At every position of the roving mass, the model frequencies of the fluid in the pipeline can be calculated. The modal frequency if is plotted will demonstrate the changes in the model frequency depending upon the positions of the roving mass. The measured signals by a roving mass of a healthy pipeline filled with air are different from the measured signals by a roving mass of a pipeline with leakage and blockage faults and can be used to detect the position and the size of leakage and blockage in the defected pipeline. It is noteworthy that the acoustic wave technique is usually determined by how severe the measurements are polluted with noise. However, signal improvement methods can be used to improve the acoustic wave reflectometry technique and the roving mass technique to effectively identify blockages and leakage in a pipeline.

Artificial intelligence has become the most effective approach which attracts many investigators to deeply research [23–66]. It has been successfully used for leak detection. The amplitude or velocity of signal propagation will change when a pipe has a leak. In [67] and [68] neural network technique has been utilized to detect the leak in a pipeline and has been provided promising results. In [69] artificial neural network has been utilized to detect the leak in a pipeline such that the sound noise data has been gathered through several microphones placed within a specific distance from the damaged part. The fast Fourier transform algorithm has been performed on data and supplied to a feed-forward network for making a final decision. In [70] neural network technique has been used for pattern recognition in oil pipe networks. After preprocessing the experimental data extracted from the acoustic sensors, it passed through a filter to produce different frequencies. The noisy data were then passed as input to the neural network. Two types of data extracted from the healthy and damaged pipe are fed to neural networks. The approach provides satisfactory results for short pipes. For long-distance pipelines, the approach is not recommended as numerous microphones are needed to place through the pipes which make the approach highly expensive. In [71] a real-time sonic leak detection system is suggested to detect leaks in oil pipelines. The wavelet transform technique has been used to extract features and also the neural network approach has been applied for making a final decision. The leak detection system is made of two digital signal processors, four piezoresistive pressure sensors, and two global positioning systems. The piezoresistive pressure sensors have been fixed at both pipe ends. The piezoresistive pressure sensors are highly susceptible to tiny variations and their measurement is based on a variation in resistance because of strain on a material. Two piezoresistive pressure sensors are applied to distinguish the signal direction during a sudden drop in pressure produced by the leak. The wavelet decomposition technique is applied to the signal obtained after preprocessing and the result is supplied as an input to the neural network. As leaks are recognized by a sudden drop in pressure, hence the cases where the pressure increases or is fixed have already been largely abandoned. However, the key technical challenge was determining an optimum sampling rate which was determined after testing some sampling rate. This technique can efficiently differentiate leakage in the pipeline and operates accordingly by turning off the pump. Nevertheless, this technique is not suitable for long-distance transportation pipeline system and it is not able to recognize the position of leakage in the pipeline. Using this technique, it is possible to monitor the system continuously. Moreover, acoustic signals sensed by sensors can successfully identify the leakage position and as well approximate the size of leakage [72]. Nevertheless, often environmental noise can generate problems in identifying the real leakage signal. This technique is not applicable for long pipelines as it requires a great number of sensors for leakage detection which is not economically beneficial.

2.4 Risk Assessment in Pipeline Failure Event

The risk corresponding to fire and explosion hazards as a consequence of leak and blockages in pipelines could not be negligent. The incidents of pipeline failure which include pipeline corrosion, human negligence could be traced back to more than fifty years ago. Pipe-failure incidents from 1959 made a quick and regulatory requirement for research and development of inspection and monitoring devices. The designed devices apart from identifying flaws, they should be able to identify the size and also the position of the leaks and blockages even under noisy environmental status. Furthermore, a tool or inspection technique that has the ability to identify a defect at its early phase is very important to the pipeline industries.

In July 1959, a fire happened in a petroleum pipeline in Vernet, Mexico, which led to the death of people and the destruction of production. In that accident at least eleven people were believed killed and forty were injured [73]. On January 17, 1962, a fire happened in a gas pipeline in Edson, Alberta which led to the death of eight people [73]. On October 12, 1965, a gas line explosion in LaSalle, Quebec, Canada demolished a number of buildings and also led to the death of twenty-eight people [74]. On the same date, a natural gas line exploded in Sundre, Alberta, Canada, and led to the death of two people [75]. In 1978 a gas line explosion in Colonia Benito, Mexico led to the death of fifty-two people [76]. On September 19, 2012, a fire happened in a natural gas pipeline in northern Mexico and led to the death of twenty-six people and injuries of forty-eight people [77, 78] as demonstrated in Fig. 2.1.

On June 6, 1989, a powerful gas pipeline explosion happened in Ufa, Russia, and destroyed a great part of the forest [78]. An investigation into the incident found that the workers by ignoring gas regulations pumped gas into a pipeline which had an undetected leakage so that caused an explosion and fire. Although there were pressure fluctuations in the gas chamber and dumping rates, the workers carried on pumping and caused a great exploration which led to the destruction of two passing trains and

Fig. 2.1 The blast of a natural gas pipe distribution centre in Reynosa, Mexico [79]

the death of nearly eight hundred people. The United States has reported the largest number of pipeline ignitions and explosions in the world as it is the world's largest oil producer with over 2.6 million miles of oil and gas pipelines. However, a majority of these accidents happen because of aging pipe systems and the limitation in the valid and efficient pipe monitoring technology. Table 2.1 demonstrates accident cases from 1959 to 2019 with the number of deaths and injuries.

The Pipeline explosions and fires are still happening up to now. The most recent accident to our knowledge has occurred in Sissonville, West Virginia [94, 95], on December 12, 2012, led to huge explosions as demonstrated in Fig. 2.2.

In the United States alone there were eighty-one accident reports for natural gas transmission and gathering pipelines in just one year [96]. However, the lack of adequate and efficient transmission line monitoring systems is one of the leading causes of pipeline accidents in the world. Even though no one died in the Sissonville pipeline accident, it resulted in seven injuries and $44 million of damage. Furthermore, in that same year in the United States, approximately seventy-one accidents of pipelines occurred which killed nine people and led to the injuring of twenty-one people. Likewise, in 2010, the explosion of the pipeline system in San Bruno, California, left eight people dead and destroyed twenty-eight houses [97]. As it was reported it took a long time to shut off the gas spewing from the pipeline in San Bruno because of the lack of automatic shut-off valves and valves that can be closed remotely. However, the reason for these failures is the lack of efficient monitoring devices for observing leakage in pipelines. Early detection and effective alarm systems will increase the time available for repairing the damage and decrease the number of fatalities. On March 12, 2014, an explosion happened in New York by a gas leak, and two buildings were destroyed, see Fig. 2.3. In that accident, eight people died and more than seventy people injured [98]. A preliminary investigation

Table 2.1 Pipeline blast with the number of deaths and injuries [80–93]

Year	Date	Region	Incident report	Amount of fatalities	Amount of injuries
1959	01/07/59	Mexico (Coatzacoalcos)	A blast of a crude oil pipeline	12	+ 100
1962	17/01/62	Canada (Edson, Alberta)	A blast of a gas lateral line	8	0
1965	05/11/65	Canada (LaSalle, Quebec)	A blast of a gas pipeline	28	0
1970	03/09/70	United States (Jacksonville)	A blast of a petroleum products pipe system	0	5
1970	09/12/70	United States (Missouri)	A blast of propane gas line	0	0
1971	17/11/71	United States (Pittsburgh)	A blast of natural gas line	6	0
1972	24/03/72	United States (Annandale, Virginia)	A blast of natural gas line	3	1
1972	14/05/72	United States (Annandale, Virginia)	A blast of crude oil line	1	2
1973	22/02/73	United States (Texas)	A blast of natural gas liquids line	0	0
1973	06/12/73	United States (Conway)	A leak in an ammonia pipe network	0	0
1974	15/03/74	United States (Farmington)	Transmission pipeline failure of Southern Union Gas Company	0	0
1974	22/04/74	United States (New York)	A blast of gas line	0	0
1974	21/05/74	United States (Texas)	A blast of natural gas gathering line	0	0
1974	06/09/74	United States (Texas)	A blast of gas line	0	0
1975	17/01/75	United States (Lima)	Crude oil terminal fire	0	0
1975	12/12/75	United States (Devers)	A blast of natural gas line	4	0
1975	02/08/75	United States (Romulus)	A blast of gas line	0	9

(continued)

Table 2.1 (continued)

Year	Date	Region	Incident report	Amount of fatalities	Amount of injuries
1976	10/01/76	United States (Fremont)	A blast of natural gas line	20	39
1976	16/06/76	United States (Los Angeles)	A blast of oil transmission line	9	14
1976	08/08/76	United States (Allentown)	A blast of natural gas transmission line	2	14
1976	09/08/76	United States (Mediapolis)	A blast of gas transmission line	6	1
1976	07/12/76	United States (Robstown, Texas)	A blast of natural gas transmission line	1	2
1977	25/01/77	United States (Williamsport)	A blast of natural gas transmission line	2	19
1977	20/07/77	United States (Creek)	A blast of gas line	2	0
1977	01/12/77	United States (Atlanta)	Rupturing of natural gas line	0	0
1977	15/12/77	United States (Lawrence)	An explosion of natural gas pipeline	2	2
1978	01/11/78	Mexico (colonia Benito Juarez)	A blast of gas line	52	11
1978	12/06/78	United States (Kansas City, Missouri)	A blast of natural gas line	0	2
1978	04/08/78	United States (Donnellson, Iowa)	A blast of gas line	2	3
1979	11/05/79	United States (Philadelphia, Pennsylvania)	A blast of natural gas line	8	19
1979	24/10/79	United States (Stanardsville, Virginia)	A blast of gas line	0	13
1986	27/10/86	Canada (Sarnia, Ontario)	A blast of gas line	0	4
2003	15/11/2003	Canada (Etobicoke, Ontario)	An explosion of the oil pipeline	7	0

(continued)

Table 2.1 (continued)

Year	Date	Region	Incident report	Amount of fatalities	Amount of injuries
2004	30/07/2004	Belgium (Ghislenghien)	A blast of natural gas line	24	122
2006	18/10/2006	Indonesia (East Java)	A blast of gas line	0	0
2010	19/12/2010	Mexico (San Martín Texmelucan de Labastida)	A blast of oil line	27	+ 50
2011	12/11/2011	Kenya (Nairobi)	A blast of the fuel pipeline	100	120
2012	18/09/2012	Mexico (Reynosa, Tamaulipas)	A blast of gas line	22	0
2013	22/11/2013	China (Huangdao, Qingdao)	A blast of oil line	55	0
2014	14/06/2014	Malaysia (Sarawak)	A blast of gas line	0	0
2014	27/06/2014	India (Andhra Pradesh)	An explosion of the gas pipeline	22	37
2017	29/04/2017	India (Jamnagar)	A blast of gas line	0	0
2019	18/01/2019	Mexico (Tlahuelilpan)	A blast of the gasoline pipeline	96	0

of the New York accident concluded the blast was because of the aging of a pipeline system and general negligence. As it was reported the pipeline was not checked for so long because of negligence.

There are fierce objections to pipeline incidents from landowners and environmental groups in the United States. Examples of these objections include the pipeline development to move crude oil from Canada to the Gulf of Mexico in 2012 [100]. The nature and extent of the pipeline was an environmental threat that drew national and presidential attention. Therefore, the safe pipeline system needed to be developed and checked before they start work.

The cumulative reported damage and fatalities caused by pipeline accidents and explosions have risen through the world. On July 17, 2010, oil from the Dalian pipeline explosion in China threatened marine animals, sea birds, and water quality as slick had spread to 430 km^2 [101]. According to the government reports the explosion resulted in a serious ecological disaster, releasing 15,000 barrels of oil into the Yellow sea [102]. Oil pipeline vandalism is a threat to Nigeria's national security. Nigeria lost approximately six to twelve billion dollars every year for the past thirty years as a result of pipeline vandalism and over 29,000 lives have been lost

Fig. 2.2 The explosion of a gas pipeline in the United States [95]

Fig. 2.3 A gas pipe system explosion in New York, United States [99]

from pipeline accidents and explosions [103]. The fire caused by pipeline vandalism in Nigeria apart from the loss of human lives it causes huge economic damage. As seen today vandals have created a major national security threat. According to the Nigerian National Petroleum Corporation on February 4, 2016, the fire caused by vandals at Arepo in Owode area of Ogun State led to the death of five pipeline security operatives [104].

Injuries and deaths due to pipeline accidents and explosions have been happened throughout the developed and developing countries all around the world [105–115]. Since the United States, Canada and Russia are three primary countries with most miles of pipelines, therefore they have the highest number of pipeline explosions. Most pipeline explosions have been in the United States, while Russia has the highest death rates in pipeline accidents. However, pipelines are regarded as the least risky way of transportation compared to trucks and tankers. They are 40 times safer than rail tanks, as well as 100 times safer than road tanks. Thus, risks associated with transmission pipelines is lower compared with other devices such as it is similar to risks associated with air travel. If a pipeline build fails, it can have catastrophic effects such as fire, explosions. Pipeline faults like leakage and blockage can result in serious ecological disasters, therefore efficient leakage and blockage awareness techniques should be developed.

2.5 The Most Common Causes of Leaking Pipes

Typically, many of the welds and protective coatings on the pipelines do not meet current safety standards and result in leakages in pipelines. Furthermore, leakages in pipelines arise from corrosion and excavation damage while pipe manufacturing. Manufacturing flaws in pipe systems may include girth or seam weld flaws by lack of fill, misalignment or, cracking. In addition, other types of leakage flaws could be corrosion at the girth welds, damage to the external pipe coating, dent stress concentration factor on the external pipe surface, and mechanical damage due to third party activities. All these factors have been described below.

2.5.1 Pipeline Damage Caused by the Stress Concentration

Cracking in pipelines can be due to stress concentration in a pipe network. It is typically associated with areas of stress concentration. For instance, polyethylene pipes are widely utilized in natural gas distribution lines and joined together. Joining is usually done by thermal fusing—heating and melting the pipe ends. Nevertheless, fusion can be the source of crack initiation such that if there exists stress concentration in the fusion area, it can result in crack initiation. In the beginning, the cracks are small but it won't take so long for a crack to grow from a certain initial size if it is not detected which could cause a leakage in the pipeline [116].

2.5.2 Pipeline Damage Caused by Third-Party Activities

Third-party damage has been the most common cause of pipeline failure which highly affects the performance of a pipeline. Third-party damage can result in pipeline leaks and ruptures. Typically, this includes mechanical damage like pipe coating damage, dent in the pipe, and gouge which may result in reduced wall thickness or distortion of pipeline cross-section. Other types of damage involve terrorist acts, sabotage, theft of the product from the pipeline, and other malevolent acts.

2.5.3 Pipeline Damage Caused by Corrosion

Fluids in pipelines influence the corrosion rate. For instance, pipelines that carry the steam around the plant and process industries, including petroleum refineries, are subject to higher rates of erosion-corrosion. Passing such fluids through steel pipelines without any internal corrosion protection can cause corrosion of the pipe wall. If left undetected, these corrosions will propagate through the wall and result in leakage. Likewise, corrosion attack upon the outside of the pipes can occur on both above-ground and underground steel pipes [117]. Often low-alloy steel pipelines have poor resistance to corrosion. Accordingly, the soil environment is responsible for stress corrosion cracking of underground stainless-steel pipelines. Stress corrosion cracking is defined as the growth of cracks under the combined influence of tensile stress and corrosive environments. There are two types of stress corrosion cracking normally developed at the surface of underground pipelines, and known as high pH stress corrosion cracking and near-neutral pH stress corrosion cracking. When these cracks appeared, they would grow in the longitudinal direction of the pipe and link up to form long shallow faults, that can lead to ruptures. In some cases, growth and interlinking of the stress-corrosion cracks produce flaws that are of sufficient size to cause leakages and subsequent pipeline ruptures. Typical growth and developments of the stress-corrosion cracks will result in subsequent failure (leakage or rupture).

2.5.4 Pipeline Damage Caused by the Operational Limitation

Depending on the circumstances, there are some operating limits for pipelines such as maximum operating pressure and the minimum pipe sizes. However, in some developing countries, there are no local regulations and laws to control the pipeline design, maintenance standards, and functions of fluid transmitting pipeline networks. A large number of pipelines have been in service for many years with no regard for any possible mechanical variations happening in the pipeline. Sometimes, some operators use the same pipelines to ship different types of hazardous liquids. However, many of the welds and protective coatings on these pipelines do not meet with the corrosive nature of the fluids. The consequences could have resulted in sever wall thickness reduction on the internal pipeline surface and pipeline rupture. Also, the pressure for the piping could exceed the pressure rating or the allowable stress for pressure design that can pass the minimum wall thickness of the pipeline. As a result, a crack could initiate which may be developed into a fracture [118].

2.6 The Most Common Causes of Blocked Pipes

Many factors influence blockage formation in pipes like the reaction of hydrocarbons with the presence of water for hydrate formation, wax deposition to the formation and eventual growth of solid layers, and deposition of suspended solid particles in the fluids (dirt, silt, clay, rust). Other factors for the formation of blockage in pipelines could be the formation of a plug, the formation of wax on pipeline walls, and other foreign materials that are produced during the crude oil refinery process. Furthermore, the inside of an aging pipe could become heavily encrusted leading to a blockage in the pipeline. All these factors have been described below.

2.6.1 Pipeline Blockage Caused by Hydrate Formation

Under the high-pressure condition, problems can arise in gas pipe systems, due to the possible presence of condensable hydrocarbons in transmission lines, particularly during extremely cold weather conditions. In ultra-deep water depth, water can enter the natural gas pipe network through the porosity of the pipe walls [119]. In ocean-bottom, the gas hydrates can be formed when water and natural gas cool in the pipes. Gas hydrates are ice-like crystalline compounds that can block pipelines in deep-sea and cause explosion in pipes [120].

In 2008 a cracked pipe caused propane fire at Valero Mckee refinery in Texas [119], see Fig. 2.4. Fire is believed to have started after a leak in the propane deasphalting unit and extended quickly, due in part to rapid fracture of the main pipe rack transferring flammable hydrocarbons. As vapour travelled in wind direction and found an ignition source, the ensuing flash fire spread. The subsequent fire injured workers and damaged unit piping and equipment [121]. Three workers suffered serious burns and several others suffered minor injuries. The fire was so large that it resulted in the evacuation and total shutdown of the McKee refinery for two months. The price of a gallon of gas went up nine cents in the west. The United States Chemical Safety Board made a safety recommendation that piping systems in refineries should be subject to formal periodic inspection and monitoring. To prevent the pipeline incidents an efficient defect inspection system, needs to be installed rather than only surface detection technique.

Fig. 2.4 Fire at Valero Mckee Refinery due to ice expansion

2.6.2 Pipeline Blockage Caused by the Agglomeration of Sand and Debris

The majority of pipe networks employed to carry crude oil or natural gas from production areas are inclined to deposit sand and other impurities, like debris. In a case that these particles could pile up for a long time, they could result in pipeline blockages. Pipeline blockages may lead to some serious problems such as interruptions in production and pipeline damages.

2.6.3 Pipeline Blockage Caused by Roots

As roots grow, they could occasionally break the pipelines and enter the cracks. When a root finds a leak, it will form root balls that block the pipeline. The chemical treatment can be used to kill plants and tree roots that have found their way into pipelines.

2.6.4 Pipeline Blockage Caused by Grease

Fats and grease are the main causes of blockages in pipe systems. They may not be harmful in liquid form but they could build up a blockage inside a pipeline if they cool.

2.7 Non-destructive Testing Methods for Leakage and Blockage Inspection

There exist many non-destructive methods recently utilized to detect leaks as well as blockages in pipe systems. The leak and blockage detection approaches are ranging from local to global techniques. The local techniques for the detection of leaks are magnetic particle inspection, visual examination, vibration monitoring, eddy current testing, dye penetrant inspection, and ultrasonic testing. The global techniques for detection of leaks are geophone and acoustic correlation techniques, pressure change monitoring, and pipeline inspection gauges.

2.7.1 Visual Inspection of Damage

Visual inspection may still be the best method for finding leaks which is walking or flying aircraft along pipelines to check for evidence of dead vegetation or wet spot around the pipeline. Furthermore, locals along the pipeline right of ways give reports of an incident to operators of the distribution systems. Although, visual inspection technique is by far the most widely used technology in the pipeline leak detection, however, its drawback is that it is unable to recognise and respond to a leak or rupture in a timely manner, also it needs physically monitoring the entire pipeline system [122].

2.7.2 Magnetic Particle Inspection of Damage

The magnetic particle inspection is a non-destructive testing technique employed for the detection of surface and near-surface defects on ferromagnetic pipe and pipe welds. The magnetic particle inspection technique induces a magnetic field on the surface of the test piece and then uses small magnetic particles to detect flaws. As a result, there appears the magnetic field over the flaw. Finally, it will clean the surface of the test piece. The magnetic particle inspection uses a 110v hand-held electromagnetic yoke magnet which is a very common piece of equipment that is utilised to establish a magnetic field. Furthermore, it uses white strippable paint as a contrasting background and a magnetic composed of iron powder particles suspended in a liquid. After that, the test area is magnetised with a yoke or permanent magnet. The defect will cause some of the lines of magnetic force to depart from the surface and form a path around the crack. Eventually, a black magnetic ink will spray onto the surface and the magnetic particles will cling to the poles and bridge the gap generated by the presence of the defect by forming a black indication against the white background. The magnetic particle inspection technique has some drawbacks which are:

- The specimen must be ferromagnetic, for example, steel and cast iron. The magnetic particle inspection method cannot be utilised for polyethylene plastic pipelines which have been widely used for gas distribution systems.
- It is necessary to remove paint or non-magnetic plating before magnetic particle testing.
- Alignment between magnetic flux and defect is important.
- The ability of this technique to detect cracks with a width of more than 2.54 mm is poor [123].
- Using this technique only small sections or small parts can be examined at one time.

2.7.3 Ultrasonic Inspection Method for Damage Detection

The ultrasonic inspection technique sends high-frequency vibrations into the part to be inspected and measures the wall thickness of the pipeline with a high level of precision and efficiency. It provides accurate thickness readings and sufficient information and detects flaws or discontinuities over the full pipe wall thickness. The ultrasonic inspection is a non-destructive testing technique that allows the sound to pass into the pipe and is ideal for testing pipes over long distances. A typical ultrasonic testing inspection system is made of various functional units, such as the pulsar and receiver, transducer, and a display device. In the ultrasonic inspection technique, the probe is placed on selected spots adjacent to the weld or area to be tested. In ultrasonic testing, the probe driven by the pulsar produces high-frequency ultrasonic energy which propagates through the materials in the form of waves to detect imperfections or to locate discontinuities in material properties. Some of the wave energy will be reflected back to the surface because of the variation in the transmission medium. Once the wave speed, as well as travel time, are known, they can later be used to identify locations of cracks or flaws.

The ultrasonic inspection is a very useful method that detects surface or near-subsurface discontinuities in pipelines. However, the drawback of this technique is that the surface of the test specimen should be reachable to transfer sound for evaluation. Another drawback associated with this technique is that a coupling medium is needed to promote the transmission of sound energy into the test specimen. Besides, irregular linear flaws that lie parallel to the beam may be missed [124]. Furthermore, the ultrasonic testing technique in the thin part of the pipe cannot perform efficiently.

2.7.4 Radiographic Technique for Damage Detection

The radiographic weld inspection is the most common non-destructive testing technique for testing fusion welds in pipes. This technique can be used to investigate the integrity of the welding structure. The technique makes the use of X-rays or gamma-rays to generate a radiographic image to check for problems. The radiation penetrates and passes through the welded pipeline linkage causing an image of the weld's internal structure being deposited on the film. The level of absorbed radiant energy by the weld depends upon the atomic number, thickness, and density of the weld. Radiant energy that has not been absorbed by the welded regions can result in the exposure of the radiographic film. That means that those regions will be too dark. Areas with minimum radiation exposure will be bright. Accordingly, cracks and voids in the interior of welds will show up as dark areas on the film [125]. Furthermore, regions of low absorption such as slag and porosity will show up as dark regions on the film and the regions of high absorption such as dense inclusions will show up as light regions on the film. Generally, radiographic testing has proven itself as a successful method of non-destructive testing for flaw inspection in weld interiors. Nevertheless, there is always a risk of damage to human cells. X-ray and gamma radiation are invisible to the naked eye and can cause severe health damage. Normally, radiation impacts health when it causes variations in the cells of the human body.

2.7.5 Pig Monitoring Systems for Damage Detection

Pigs have been mainly utilised in oil and gas pipelines for pipeline cleaning. Initially, hydro-testing and pigs are utilised to fill the line with water. After the hydrostatic test has been made, it is important to remove all water used for hydrostatic test and place the line in service. The pig is usually equipped with sensors and data recording devices and can be used at regular intervals to check the internal conditions of the pipes [126]. Large quantities of data can be built up over a period of time by smart pigs to provide a history of the line and identify changes in pipeline conditions over time. Pigs can be introduced into the line to remove the deposits via pig traps [127]. Normally, pigs are constructed and operated so that sealing elements provide a positive interference with the pipeline walls. Generally, pigs are inserted into a pipeline and move in the direction of the applied force (pressure). Higher differential pressure between the tail and nose of the pig is needed to start the pig. Furthermore, for maintaining a constant motion, appropriate flow speed is needed.

Inspection pigs can also be used for many different reasons in pipelines. In pipeline transportation, the pigging system starts with the installation of a pig trap that includes the pigs, the pig launcher, and the pig receiver. The pig can be installed without stopping the flow in the pipeline, and it can be forced through the line by either product flow, or by being towed by a cable or other device. The pipeline pig is typically a spherical or cylindrical device that sweeps the distributing inhibitor throughout the inner walls of the pipeline by scraping the sides of the pipeline and pushing debris towards the exit point. As pigs travel along the line, apart from cleaning the pipeline, they can further assist inspect the inner pipeline condition to prevent and identify problems.

The very first pigs were constructed from straw bales wrapped in barbed wire and were created to clean process lines. There lies the two basic technology vision for why the process of the maintenance is named pipeline pigging, even though neither has been proved. One theory is that pig is derived from the initial letters of the term Pipeline Intervention Gadget or Pipeline Inspection Gauge. The other states that in the past, pipeline pigs were inserted into the line with leather sealing discs, which squeaked against the pipeline sides making it sound similar to the squealing of a pig [128]. Pigs can provide useful information about corrosion, laminations, leaks, and ruptures in pipes.

Operators should consider a number of factors while choosing a suitable pig for cleaning and maintenance of the pipelines. Initially, it is essential to set the objective and define the task that the pig has to perform. Moreover, pipe diameter and operating pressures and temperatures with regard to pig type are necessary. Also, the piping design is integral when selecting the appropriate pig. Since each piping system is different, there exists no predetermined schedule for smart pigging the system. However, pigging frequencies are based on the amount of debris and the amount of wear and tear on the pipeline. The simple pig is propelled by the fluid in the pipeline and collects information and once extracted from the pipe the data are analysed. A proper pig receiver acts as a point to remove the pig from the pipeline as a process of collection and subsequent analysis of data.

There are several basic technologies most commonly used for gathering pipeline information applying smart pigs. Common techniques include magnetic flux leakage detection technique and ultrasonic inspection technique. Magnetic flux leakage is a magnetic technique of nondestructive testing that is based on magnetizing the pipe wall. Any loss of wall thickness (for example, due to corrosion) will result in a change of the magnetic field which is later recorded by a magnetic field sensor linked to the pig. The signal that is detected and recorded by smart pig is assessed based on reference signals to identify any flaw in the pipeline wall. Many pigs use the ultrasonic method to transfer ultrasonic pulses into the pipe wall and receive a reflected ultrasonic signal that is reflected by a defect in the pipe. The reflected sound from the inside and outside tube wall to calculate the wall thickness identifies areas of damage in the pipeline and depends on the speed of the pig. Intelligent pigs can indicate where leaks or other damage has affected the pipeline wall. After a pipeline is constructed and before it is put into service it is usually pigged and risk assessment information about the pipeline is assessed to recognize threats to the pipeline and evaluate risk. This inspection methodology is known as zero or baseline pigging.

In pipeline transportation, it is necessary to know that the pipe is piggable. Also, it is essential to know that there exist no physical obstacles in the piping network that makes the passage too tight and there also exist smart pigs for the aim of identifying leaks and damages in the pipeline.

2.7.6 Boiling Water Reactor for Damage Detection

In operating nuclear power plants, piping systems are subjected to various operating and environmental stressors which could cause ruptures in the pipeline [129, 130]. Boiling water reactor safety systems are nuclear safety systems constructed within boiling water and operate as fault detection systems. The boiling water reactor online monitoring system uses the evaluation of two signals, periodic and random fluctuations, captured from a measurement system placed on a nuclear reactor. The method is based on a prime-factor algorithm utilized for calculation of the fast Fourier transform. Thus, the reference calculation signals are compared to power spectral density functions obtained by applying a fast Fourier transform method [130].

2.7.7 Adding an Odourant to the Fluid for Damage Detection

The most cost-effective technique to detect leakage in pipelines is to simply add an odourant to the fluid which could locate leaks. This technology makes it possible to detect leaks using trained sniffer dogs. When the odourant is added to natural gas, it could aid the detection of leakages. Organic compounds are used for odorization and make the most effective odorizers for odourless fluid. For instance, methane is a colorless, odourless gas that doesn't smell. The largest component of natural gas is formed by methane. Since natural gas doesn't actually smell like anything, the harmless artificial smell is added by the gas company so that people can detect gas leaks with their noses. Harmless chemicals like mercaptan or Trimethylamine that are smelling like rotting eggs and rotting fish can be added in fewer amounts to give gas a distinctive odour and allow the detection of leakages without any external equipment. Nevertheless, in case of leakage happens in a region with no folk, leakage along the pipe will probably go undetected for an indefinite period of time except in the case, the pipeline system is watched by line detection group using proper equipment, e.g. gas sniffer. Also, the price of inspections and maintenance checks for leaks in pipelines are really high. Moreover, leakages in the transmission system are not detectable if the odourant concentration is blown in the opposite direction by severe weather status including summer storms and hurricanes thus this can become problematic when locating the leakage. Furthermore, the odour of the gas could be filtered out while moving from a landfill through the soil into the outdoor air. If the gas passes through sewer containing due to strong and offensive odour of other combustible fluid your nose can become overwhelmed by other smells and you cannot smell the gas [131]. Although, adding an odourant to the fluid for damage detection is considered as a common way, however, it is not an efficient method for identifying leaks from natural gas pipelines.

2.7.8 *Mass-Volume Balance Technique for Damage Detection*

The operating principles of the primary leakage inspection technique were based upon the mass balance, that used the concept of conservation of mass [132–136]. The mass-volume balance method for leakage detection is based on the fact that under the steady-state condition and in the absence of leak, the inflow and outflow measured at the two ends of the pipeline section must be balanced and any changes in flow rate and pressure may present the existence of a leakage. Flow measurement technique depends on observing flow meters or gauges at periodic intervals. Ideally, local monitoring of flow needs no telemetry. The law of conservation of mass or principle of mass conservation states that for any system the mass balance approach performs similarly with volume balance approach [50]. This means that when a fluid flows to the pipeline section may stay in the pipeline section or depart to another section. In standard piping network estimated flow rates can be provided for flows leaving or entering the network. In general, if the measurement error is less than the threshold value, there is no leakage and if it is greater than the established threshold value the leakage could be detected [72]. This method has been commercialised and is widely adopted in the oil industry. This technique has a high degree of sensitivity to pipe instrumentation precision. However, in this technique, the use of the steady-state assumption may lead to the increment of the inspection period for preventing false alarms and hence resulting in delaying the response time to leakages. For example, a 1% leakage requires nearly 40–60 min to identify [72]. The major drawback of the mass balance technique is its inability to locate the position of leakage. Therefore, other techniques should be used along with the mass balance technique after the leakage has been inspected to locate the position of the leakage.

There are several disadvantages to using the mass balance approach [136]. A mass balance technique is only possible under steady-state conditions. It has limited implementation when dealing with gas pipe systems. In order to make the technique feasible, precise measurement of flow on both sides of the pipe is essential [137]. This approach could be more efficient for pipelines if the flow measuring device precisely measures the mass entering and leaving the piping system. Orifice meters are the most common meters utilised for fluid flow measurement and have nearly 2–4% precision [138]. Hence, the small variation in measurement may cause abnormal outcomes. As a result, the inaccurate measurement could lead to unnecessary costs and shutting down the piping system for pressure testing.

2.7.9 *Real Time Transient Technique for Damage Detection*

The real-time transient modeling method utilises the conservation principle of momentum and energy to create mathematical models of the flow within the pipeline [139]. The real-time transient modeling technique is not capable of providing rapid response time as well as accurate results for the position of the leak in the pipeline [140]. Experts are needed continually for adjusting the mathematical model to compensate for alterations in the physical properties of the pipeline.

Real-time transient modeling is a pipeline leak detection technique that is formed by mass, momentum, and energy conservation equations used to model the flow. It is only in the presence of a leakage that the measured flow value and the estimated flow value will diverge. Evaluations of the flow, temperature, and pressure at the two ends of the line are essential to construct the real-time transient model. Since noise can cause false alarms and impact the reliability of the technique therefore, continuous monitoring of the system and noise level will keep fire risk to a minimum.

In [141], a real-time transient modeling method has been utilised for leakage detection and localization in the pipeline systems. In that work, they used mathematical dynamic models, nonlinear adaptive state observers as well as a correlation inspection method. Their technique was examined for a gasoline piping system of 68 km length [142]. The Experiments indicated that inspection of the leak having the size of 0.2% of input flow is possible within 90s. Furthermore, in that study, the position of leakage was approximated with a precision of 0.9%. In [143–147] a linearized model of the pipeline flow is utilised on the model by a network of N nodes. Pressure, as well as flow rate, are used for the evaluation at the two ends of the line. As a set of partial differential equations is used for modeling fluids transported in pipeline, therefore, a finite dimension nonlinear model is obtained utilising measured pressures as a function of the input. The approximated flow value represents the model output. In [143] an extended version of a real-time transient modeling method to estimate two leaks simultaneously in a piping system is proposed. The authors in [148] focused on leakage reconstruction in pipe systems utilising sliding mode observer. Single and multiple leakage situations are considered in that study. The single and multiple leakage locations are specified using an efficient method in a finite duration of time and under necessary and sufficient status. In [149] a leak inspection device consisting of an adaptive Luenberger-type observer based upon a set of two-coupled partial differential equations governing the flow dynamics is proposed. This technique has the advantages of high accuracy in identifying very small leakages and also approximating the size of them. The main limitation of this method is that it deals with a large data set in real-time which is highly costly [72]. A modified version of that technique has been developed in [150].

2.7.10 Supervisory Controls and Data Acquisition System for Damage Detection

The supervisory controls and data acquisition system is one of the most commonly used types of industrial control systems that is utilised to observe piping systems at most block valve's status. The supervisory controls and data acquisition system processes information collected from sensors throughout the pipe system. Local remote terminal units are responsible for maintaining information gathered from sensors. The real-time information collected at the local remote terminal units is later moved to a central station using cellular phone or satellite connections. The transferred data are collected and automatically processed by the main control room to the pipeline operators through a set of screens or human–machine interface software which enables operators to monitor all system operations. The data allow operators to see the complete picture of the pipeline's condition. The difference between the model calculated and the actual measured values can represent leakages. The supervisory controls and data acquisition technique tries to mathematically model the hydraulic behaviour of the pipeline. The origin of the hydraulic equations are the conservation laws for mass, momentum, and energy combined with the equation of state that is used to relate the density of a fluid to pressure as well as temperature change. These equations are useful for output pressure and flow rate calculations utilizing the supplied input parameters of the piping system. The validity of the simulation results then had to be compared to the calculated results. A relatively significant discrepancy between the simulation results and calculated results indicate that there exists damage in the pipe. The conservation rules utilized in the supervisory controls and data acquisition system can be written as a non-steady partial differential equation when hydraulic parameters such as temperature as well as flow are considered as direct functions of time and distance through the line. This technique has the advantages of high accuracy in identifying leakages and also approximating the size of them. Also, large leakages can be identified more quickly and located more accurately with this technique [151]. The main limitation of this method is that:

1. It can identify tiny leakages over a very long time,
2. It is not able to identify leakages from a pipe system.

2.7.11 Acoustic Emission Technique for Damage Detection

The acoustic emission technique uses noise or vibration generated as a result of a sudden drop in pressure to identify the incidence of a pipe leak. The appearance of leakage is mainly due to increases in noise level. This is due to turbulence at high speeds. Normally, leakage sound is usually within the frequency of 1 kHz–1 MHz. The noise of fluid escaping a pipe generates various turbulent properties from the environmental effects and is employed to identify the leakage and its location in pipelines. Leakage voice usually propagates at a much slower speed in the pipe wall than in a fluid-filled pipe. The leakage voice can be identified using sensors that are installed in the pipeline. The lowest level detectable leakage ratio using the acoustic emission method is in the range of 5–10 m^3s^{-1} for gas, and 6–10 m^3s^1 for liquid. The main drawbacks of this method are that [152, 151]:

1. high-frequency signals will be attenuated more in the tiny diameter pipelines, therefore needs short distance between sensors
2. signal attenuation in pipes could vary due to various factors such as soil or depths that could complicate the real position of acoustic emission source.

The acoustic pressure wave technique examines the rarefaction waves generated once a leakage happens. In general, this technique generates an acoustic noise around the place of a leak such that the obtained signal is stronger around the leak site hence enabling leak location. When the hole forms, fluid escapes in the shape of a high-speed jet and leads to creating a low-frequency acoustic signal that could be identified. When these signal properties are different from the base line, alarms will be activated immediately by the alarm generator [153]. As a result, negative pressure waves will be generated that may grow in both directions along the pipe and could be identified. The acoustic emission technique operates based on the significant feature of pressure waves moving over large distances at the velocity of sound directed by the pipe walls. Normally, the obtained signal is stronger around the leak site, therefore, assist to identify the location of leakage. The cross-correlation procedure is the most widely used technique in acoustic approaches to detect and localize leaks in pipelines. This technique finds the location of a leakage in a pipeline from the reflection of a pressure wave from the leakage such that the increment in the leak size in pipelines give rise to pressure waves. More accurate leak detection is carried by analysing data from pressure sensors using complex mathematical models. Nevertheless, the technique fails to find an ongoing leakage after the first incident. After a break or fracture in the pipe wall, there will not be any subsequent pressure waves in the pipe. Hence, this technique is not able to identify any ongoing leakages.

2.7.12 *Acoustic Pulse Reflectometry Technique for Damage Detection*

The acoustic pulse reflectometry method has been used successfully to identify damage in pipelines utilising the time domain [137, 154]. In [21], this method has been employed to identify faults at a distance of 5 km from the acoustic source. Applying the acoustic pulse reflectometry method, a small pulse is injected in a given pipe system and the reflections produced are recorded by sensors. Mainly, there are two types of evaluations; one without showing weakness or flaws in the pipeline and another with weaknesses or flaws. Discontinuity in a piping system, like a leak or blockage, can lead to the impedance variation. The acoustic pulse reflectometry is an efficient technique to observe the health of a pipe. Nevertheless, the presence of measurement noise can seriously affect the estimation of this technique such that makes accurate measurement results difficult to achieve. A review of previous works on the acoustic pulse reflectometry approach shows that this approach has successfully been applied to identify faults in pipelines but not without measurement noise.

2.8 Signal Processing Methods for Damage Identification

A more realistic denoising technique is required for improving the capabilities of the acoustic pulse reflectometry approach. Additionally, a combination of a signal processing approach with an acoustic wave propagation approach can also be useful for exact fault detection in pipelines. There exist various signal processing methods recently utilised for signal decomposition and denoising such as fast Fourier transform technique, cepstral analysis technique, and wavelet transform technique.

2.8.1 Cepstral Analysis Technique for Damage Identification

The cepstrum analysis technique is described as the inverse Fourier transform of the natural logarithm of the power spectrum of the signal with linear frequency [155]. It can be utilised for the identification of any periodic structure in the logarithmic spectrum, for instance, the multiples of the fundamental frequencies. The cepstrum analysis technique is sensitive to the identification of the presence of echo. Furthermore, this technique has the ability to clearly split up specific source and transmission path impacts of a deconvolution. There exists a simple relation between the input signal and output signal in a stable system and could well be described in the form of the frequency domain, complex spectra, or in the form of the power spectra. Due to the linearity property of the inverse Fourier transform, the maintainer relation can be retained in the cepstrum. In [156] the cepstrum analysis technique is utilised to identify leaks in pipes. In [157] the cepstrum analysis approach is used to identify leakages in pipelines to derive data information on the time delay between the developed wave and obtained echo results in the piping systems.

2.8.2 Fast Fourier Transform Technique for Damage Identification

The Fourier transform of a time-domain signal can be defined as frequency-amplitude that provides the frequency domain representation of the original signal. Therefore, the frequency value comes from the signal generator such that each frequency component can be assigned to a corresponding amplitude value. Discrete Fourier transform is a method to convert time domain data into frequency domain data. This operation is useful in many fields, however, it requires many sum and multiplication operations over complex exponentials, and leads to growing the calculation time. For computing the discrete Fourier transform with reduced execution time, a fast Fourier transform technique is introduced [158]. Fast Fourier transform technique is one of the most efficient approaches which shows promising result in determining the position of the leakage [159–161]. Although fast Fourier transform for time–frequency analysis is still broadly utilised in industry, it cannot be an appropriate method for a non-stationary signal. Therefore, if the spectral characteristics of the analysed signal are time-varying, a transform approach that gives the time–frequency representation of a signal is required. Time–frequency transforms such as the wavelet transform has proven to be an efficient algorithm.

2.8.3 Wavelet Transform Technique for Damage Identification

The wavelet transform technique has become a very effective tool in signal processing and the exact reconstruction of signals. This technique has the ability to execute local analysis of a signal, for example zooming on any desired interval of space or time. Recent research efforts [162–171] have established the potential of wavelet transform technique for signal decomposition and reconstruction of the signal to obtain a de-noised signal in damage inspection in the structures [161, 162], enhancing the image quality [172, 173], damage identification in rotating machinery [174, 175], fibre-optic, medical field, and muscle fatigue identification [176, 177]. Wavelet transform techniques that have been recently utilised for signal analysis are the continuous wavelet transform, discrete wavelet transform and stationary wavelet transform. Continuous wavelet transform utilises inner products to evaluate the similarity between a signal and an analysing function. It is obtained by continuously shifting the scale of the analysis window function, translating the window function in time, multiplying by the signal, and integrating the result overall time. In the discrete case, the signal to be examined is passed through filters with various cutoff frequencies at different scales. The signal is estimated by passing it through a series of low pass and high pass filters. After each filtering stage, the filter outputs are down-sampled by 2 as half of the samples are removed after passing the signal through a lowpass filter. Down-sampling is the process of decreasing the sample rate by an integer factor, for example, down-sampling by a factor of 2, that means you simply "throw away" every second sample and down-sampling by a factor n means we only keep every nth sample and discard every sample in between [177]. In numerical analysis and functional analysis, a discrete wavelet transform is a non-redundant decomposition analysis that generates a nonredundant representation of the signal and it needs a significant amount of calculation time to calculate the discrete wavelet transform coefficients under each scale factor [161] which leads to an algorithm that is performing poorly in the presence of noise. Stationary wavelet transform is an up-sampling procedure that can considerably eliminate or suppress the noise and maintain the feature frequency of the vibration signal [178–180]. Stationary wavelet transform procedure can be carried out before performing filter convolution at each scale. It is non- invariant in time and space such that the coefficients of a delayed signal are not a time-shifted version of those of the original signal [181].

In [182] the de-noising performances of three algorithms, continuous wavelet transform, discrete wavelet transform, and stationary wavelet transform are analysed and compared. The comparison results show that continuous wavelet transform and stationary wavelet transform yield better results than the discrete wavelet transform. In [183] the stationary wavelet transform approach is used for processing ultrasonic signals and verify its effectiveness. A hole is made on a flat bottom and an ultrasonic inspection is performed. The results suggest that the technique has relatively low noise. Also, the experimental results showed that the stationary wavelet transform has the advantage of information integrity and time translation invariance after signal de-noising. Furthermore, the results are reliable indicators of precise defect localization. In [184] a new method based on auxiliary mass spatial probing by the stationary wavelet transform is suggested to detect damage in beams. In their work, the mode shape of damage beams is decomposed by stationary wavelet transform into a smooth curve, called the approximation coefficient, and a detail coefficient curve. Following the decomposition, the detail coefficient curves are de-noised. The new de-noised detail coefficient and the detail coefficient curves have useful information for damage inspection and therefore have been utilised to identify flaws in beam-like structures. In [185] the stationary wavelet transform technique is used to identify flaws in rotating motor bearings. Healthy motor bearings are compared to the faulty case using stationary wavelet transform technique such that the considered fault for the comparison was a bearing fault. The comparison results show that stationary wavelet transform yields better results than the discrete wavelet transform. In [186] a combination of the stationary wavelet transform, Kurtosis, and cross-correlation algorithms is proposed to effectively improve the ultrasonic signal detection ability. They also used the Meyer wavelet to obtain pulse responses from ultrasonic signals. Some recent developments of the stationary wavelet transform technique can be found in [173, 187–198].

References

1. U.S Department of Transportation FAA.: Flight Standards Service-Aviation Maintenance Technical Handbook-Airframe, 2 (12 and 14) (2012). doi: FAA-H-8083-31
2. Folga, S.M.: Natural Gas Pipeline Technology Overview.: Argonne National Laboratory ANL/EVS/TM/08-5:1-12 (2007)
3. Chris, T., Saguna, A.: Pipeline Leak Science Series. 5th Tome 1st Fasc (2007)
4. Barker, M., Fessler, R.R., Biztek, R.: Pipeline and Hazardous Materials Safety Administration Office of Pipeline and Safety, Integrity Management Program, Pipeline Corrosion. Final Report, US Department of Transportation, Under Delivery Order DTRS56–02-D-70036 (2008)
5. Machell, J., Mounce, S., Boxall, J.B.: Online Modelling of Water Distribution Systems: A UK Case Study, Drinking Water Engineering and Science (2010)
6. Sider, A.: High-tech monitors often miss oil pipeline leaks. Wall Street J. (2014)
7. Warda, H.A., Adam, G., Rashad, A.B.: A practical implementation of pressure transient analysis in leak localization in pipelines. Int. Pipeline Conf. (2004). https://doi.org/10.1115/IPC2004-0551

8. Baker, M., Fessler, R.R., Biztek, R.: Pipeline Corrosion. Final Report, U.S. Department of Transportation, Pipeline and Hazardous Materials Safety Administration Office of Pipeline Safety. Integrity Management Program, Under Delivery Order (2008)
9. Willcox, M., Downes, G.: A Brief Description of NDT Techniques. Insight NDT Limited (2000–2003)
10. Duan, H.F., Lee, P.J.: Experimental investigation of wave scattering effect of pipe blockages on transient analysis. In: 16th Conference on Water Distribution System Analysis 891314-1320WDSA Science Direct (2014)
11. Lazhar, A., Hadj-Taieb, L., Hadj-Taieb, E.: Two leaks detection in viscoelastic pipeline systems by means of transient. J. Loss Prev. Process Ind. **26**, 1341–1351 (2013)
12. Duan, H.F., Lee, P., Ghidaoui, M.: Transient wave-blockage interaction in pressurized water pipelines. In: 12th International Conference on Computing and Control for the Water Industry, CCW12013 Science Direct, vol. 70, pp. 573–582 (2014)
13. Kaliatka, A., Vaisnoras, M., Valincius, M.: Modelling of valve induced water hammer phenomena in a district heating system. J. Comput. Fluids **94**, 30–36 (2014)
14. Riedelmeier, S., Becker, S., Schlucker, E.: Measurements of junction coupling during water hammer in piping systems. J. Fluids Struct. **48**, 156–168 (2014)
15. Tijsseling, A.S.: Water hammer with fluid-structure interaction in thick-walled pipes. Comput. Struct. **85**, 844–851 (2007)
16. Wang, R., Wang, Z., Wang, X., Yang, H., Sun, J.: Water hammer assessment techniques for water distribution systems. Procedia Eng. **70**, 1717–1725 (2014)
17. Delgado, J.N., Martins, N.M.C., Covas, D.I.C.: Uncertainties in hydraulic transient modelling in raising pipe systems. Laboratory case studies Science Direct Procedia Engineering, vol. 70, pp. 487–496 (2014)
18. Nayak, C.: Fault detection in fluid flowing pipes using acoustic method. Int. J. Appl. Eng. Res. **9**(1), 23–28 (2014)
19. Sharp, D.B., Campbell, D.M.: Leak detection in pipes using acoustic pulse reflectometry. Acustica **83**(3), 560–566 (1997)
20. Duan, W., Kiby, R., Prisutova, J., Horoshenkov, K.V.: On the use of power reflection ratio and phase change to determine the geometry of blockage in a pipe. J. Appl. Acoust. **87**, 190–197 (2015)
21. Hunaidi, O., Wang, A.: A new system for locating leaks in urban water distribution pipes. Int. J. Manage. Environ. Q. **17**(4), 450–466 (2006)
22. Duan, H.F., Lee, P.J., Ghidaoui, M.S., Tuck, J.: Transient wave-blockage interaction and extended blockage detection in elastic water pipelines. J. Fluids Struct. **46**, 2–16 (2014)
23. Jafarian, A., Jafari, R.: Approximate solutions of dual fuzzy polynomials by feed-back neural networks. J. Soft Comput. Appl. **2012**, 1–5 (2012)
24. Jafarian, A., Jafari, R.: An iterative method for solving fuzzy polynomials by fuzzy neural networks. In: 6th International Conference on Fuzzy Information and Engineering, Iran (2012)
25. Jafarian, A., Jafari, R.: Simulation and evaluation of fuzzy polynomials by feed-back neural networks. In: 6th International Conference on Fuzzy Information and Engineering, Iran (2012)
26. Jafarian, A., Jafari, R.: New iterative approach for solving fully fuzzy polynomials. Int. J. Fuzzy Math. Syst. 3 (1) (2013)
27. Jafarian, A., Jafari, R.: New method for solving fuzzy polynomials. Adv. Fuzzy Math. **8**(1), 25–33 (2013)
28. Jafari, R., Yu, W.: Uncertainty nonlinear systems control with fuzzy equations. In: 2015 IEEE International Conference on Systems, Man, and Cybernetics, pp. 2885–2890. IEEE (2015)
29. Jafari, R., Yu, W.: Artificial neural network approach for solving strongly degenerate parabolic and burgers-fisher equations. In: 2015 12th International Conference on Electrical Engineering, Computing Science and Automatic Control (CCE), pp. 1–6. IEEE (2015)
30. Jafari, R., Yu, W.: Fuzzy control for uncertainty nonlinear systems with dual fuzzy equations. J. Intell. Fuzzy Syst. **29**(3), 1229–1240 (2015)
31. Jafarian, A., Jafari, R., Golmankhaneh, A.K., Baleanu, D.: Solving fully fuzzy polynomials using feed-back neural networks. Int. J. Comput. Math. **92**(4), 742–755 (2015)

32. Jafari, R., Yu, W., Li, X.: Solving fuzzy differential equation with Bernstein neural networks. In: 2016 IEEE International Conference on Systems, Man, and Cybernetics, 2016. vol 978-1-5090-1897-0/16/$31.00 ©2016 IEEE. 2016 IEEE International Conference on Systems, Man, and Cybernetics • SMC, pp. 1245–1250 (2016)
33. Jafarian, A., Jafari, R.: A new iterative approach based on artificial intelligence for solving dual fuzzy polynomials. In: 2nd International Conference on Technology-Engineering and Science (2016)
34. Jafarian, A., Jafari, R., Al Qurashi, M.M., Baleanu, D.: A novel computational approach to approximate fuzzy interpolation polynomials. SpringerPlus **5**(1), 1428 (2016)
35. Razvarz, S., Jafari, R.: Experimental study of Al_2O_3 nanofluids on the thermal efficiency of curved heat pipe at different tilt angle. In: 2nd International Conference on Technology-Engineering and Science (2016)
36. Jafari, R., Yu, W.: Fuzzy Modeling for Uncertainty Nonlinear Systems with Fuzzy Equations. Mathematical Problems in Engineering 2017 (2017) https://doi.org/10.1155/2017/8594738
37. Jafari, R., Yu, W.: Uncertain nonlinear system control with fuzzy differential equations and Z-numbers. In: 2017 IEEE International Conference on Industrial Technology (ICIT), pp. 890–895. IEEE (2017)
38. Jafari, R., Yu, W., Li, X.: Fuzzy differential equations for nonlinear system modeling with Bernstein neural networks. IEEE Access **4**, 9428–9436 (2017)
39. Jafari, R., Yu, W., Li, X.: Numerical solution of fuzzy equations with Z-numbers using neural networks. Intell. Autom. Soft Comput., 1–7 (2017)
40. Razvarz, S., Jafari, R.: Intelligent techniques for photocatalytic removal of pollution in wastewater. J. Electr. Eng. **5**(1), 321–328 (2017a)
41. Jafari, R., Razvarz, S., Gegov, A.: A New Computational Method for Solving Fully Fuzzy Nonlinear Systems. In: 10th International Conference Computational Collective Intelligence, ICCCI 2018, Bristol, UK, 5–7 Sept. 2018
42. Jafari, R., Razvarz, S., Gegov, A.: A novel technique to solve fully fuzzy nonlinear matrix equations. In: 13th International Conference on Applications of Fuzzy Systems and Soft Computing: ICAFS 2018. Springer (2018)
43. Jafari, R., Razvarz, S., Gegov, A.: Solving differential equations with z-numbers by utilizing fuzzy Sumudu transform. In: Intelligent Systems and Applications, vol 869, pp. 1125–1138. Springer International Publishing (2018)
44. Paul, S., Yu, W., Jafari, R.: A method for bidirectional active vibration control of structure using discrete-time sliding mode. IFAC-PapersOnLine **51**(13), 361–365 (2018)
45. Jafari, R., Razvarz, S., Gegov, A.: End-to-end memory networks: a survey. In: Advances in Intelligent Systems and Computing. Springer (2019)
46. Jafari, R., Razvarz, S., Gegov, A., Vatchova, B.: A survey on applications of neuro-fuzzy models. In: Proceedings of the 10th IEEE International Conference on Intelligent Systems, 2019. Institute of Electrical and Electronics Engineers (2019)
47. Jafari, R., Razvarz, S., Yu, W., Gegov, A., Goodwin, M., Adda, M.: Genetic algorithm modeling for photocatalytic elimination of impurity in wastewater. In: Proceedings of SAI Intelligent Systems Conference, pp 228–236. Springer, Cham (2019)
48. Jafari, R., Razvarz, S., Gegov, A.: A novel technique for solving fully fuzzy nonlinear systems based on neural networks. Vietnam J. Comput. Sci. (2020)
49. Jafari, R., Razvarz, S., Gegov, A., Vatchova, B.: Deep learning for pipeline damage detection: an overview of the concepts and a survey of the state-of-the-art. In: Proceedings of the 10th IEEE International Conference on Intelligent Systems. Institute of Electrical and Electronics Engineers (2020)
50. Jafari, R., Razvarz, S., Vargas-Jarillo, C., Gegov, A.: Blockage detection in pipeline based on the extended Kalman Filter observer. Electronics **9**(1), 91 (2020)
51. Paul, S., Yu, W., Jafari, R.: Stability Analysis and Bidirectional Vibration Control of Structure. In: Emerging Trends in Civil Engineering, pp. 275–287. Springer, Singapore (2020)
52. Jafarian, A., Jafari, R.: A new computational method for solving fully fuzzy nonlinear matrix equations. Int. J. Fuzzy Comp. Modell. **2**(4), 275–285 (2019)

53. Jafari, R., Razvarz, S.: Solution of fuzzy differential equations using fuzzy Sumudu transforms. In: 2017 IEEE International Conference on INnovations in Intelligent SysTems and Applications (INISTA), 3–5 July 2017, pp. 84–89 (2017). https://doi.org/10.1109/INISTA.2017.8001137
54. Jafari, R., Yu, W., Li, X., Razvarz, S.: Numerical solution of fuzzy differential equations with Z-numbers using Bernstein neural networks. Int. J. Comput. Intell. Syst. **10**(1), 1226–1237 (2017)
55. Razvarz, S., Jafari, R.: ICA and ANN modeling for photocatalytic removal of pollution in wastewater. Math. Comput. Appl. **22**(3), 38 (2017b)
56. Razvarz, S., Jafari, R., Yu, W., Golmankhaneh, A.K.: PSO and NN modeling for photocatalytic removal of pollution in wastewater. In: 2017 14th International Conference on Electrical Engineering, Computing Science and Automatic Control (CCE), 20–22 Oct. 2017, pp 1–6 (2017). https://doi.org/10.1109/ICEEE.2017.8108825
57. Jafari, R., Razvarz, S., Gegov, A.: A new computational method for solving fully fuzzy nonlinear systems. In: International Conference on Computational Collective Intelligence, pp. 503–512. Springer (2018)
58. Jafari, R., Razvarz, S., Gegov, A.: Solving differential equations with Z-numbers by utilizing fuzzy sumudu transform. In: Proceedings of SAI Intelligent Systems Conference, pp. 1125–1138. Springer, Cham (2018)
59. Jafari, R., Razvarz, S., Gegov, A.: Fuzzy differential equations for modeling and control of fuzzy systems. In: International Conference on Theory and Applications of Fuzzy Systems and Soft Computing, pp. 732–740. Springer, Cham (2018)
60. Jafari, R., Razvarz, S., Gegov, A., Paul, S.: Modeling and control of uncertain nonlinear systems. In: 2018 International Conference on Intelligent Systems (IS), 25–27 Sept. 2018, pp 168–173 (2018). https://doi.org/10.1109/IS.2018.8710463
61. Jafari, R., Razvarz, S., Gegov, A.: Paul S Fuzzy modeling for uncertain nonlinear systems using fuzzy equations and Z-numbers. In: UK Workshop on Computational Intelligence, pp. 96–107. Springer (2018)
62. Razvarz, S., Jafari, R., Gegov, A., Paul, S.: Neural network approach to solving fully fuzzy nonlinear systems. In: Fuzzy Modeling and Control: Methods, Applications and Research, pp. 46–68. Nova Science Publishers, Inc. (2018)
63. Razvarz, S., Jafari, R., Granmo, O.-C., Gegov, A.: Solution of dual fuzzy equations using a new iterative method. In: Asian Conference on Intelligent Information and Database Systems, pp. 245–255. Springer (2018)
64. Razvarz, S., Jafari, R., Yu, W.: Numerical solution of fuzzy differential equations with Z-numbers using fuzzy Sumudu transforms. Adv. Sci. Technol. Eng. Syst. J. (ASTESJ) **3**, 66–75 (2018)
65. Jafari, R., Razvarz, S., Gegov, A.: Neural Network Approach to Solving Fuzzy Nonlinear Equations using Z-Numbers. IEEE Trans. Fuzzy Syst., 1 (2019).https://doi.org/10.1109/TFUZZ.2019.2940919
66. Jafari, R., Razvarz, S., Gegov, A., Paul, S., Keshtkar, S.: Fuzzy Sumudu transform approach to solving fuzzy differential equations with Z-numbers. In: Advanced Fuzzy Logic Approaches in Engineering Science, pp. 18–48. IGI Global (2019)
67. Belsito, S., Lombardi, P., Andreussi, P., Banerjee, S.: Leak detection in liquefied gas pipelines by artificial neural networks. AIChE J. **44**(12), 2675–2688 (1998)
68. Bicharra, A.C., Ferraz, I.N., Bernardini, F.C.: Artificial neural networks ensemble used for pipeline leak detection systems. In: Proceeding of the Seventh International Pipeline Conference, Alberta, Canada, Sept. 29–Oct. 3 2008. Citeseer (2008)
69. Shibata, A., Konishi, M., Abe, Y., Hasegawa, R., Watanabe, M., Kamijo, H.: Neuro based classification of gas leakage sounds in pipeline. In: 2009 International Conference on Networking, Sensing and Control, pp. 298–302. IEEE (2009)
70. Zhao, J., Li, D., Qi, H., Sun, F., An, R.: The fault diagnosis method of pipeline leakage based on neural network. In: 2010 International Conference on Computer, Mechatronics, Control and Electronic Engineering, pp. 322–325. IEEE (2010)

71. Avelino AM, de Paiva JA, da Silva RE, de Araujo GJ, de Azevedo FM, Quintaes FdO, Maitelli AL, Neto AD, Salazar AO Real time leak detection system applied to oil pipelines using sonic technology and neural networks. In: 2009 35th Annual Conference of IEEE Industrial Electronics, 2009. IEEE, pp 2109–2114
72. Murvay, P.S., Silea, I.: A survey on gas leak detection and localization techniques. J. Loss Prev. Process Ind. **25**(6), 966–973 (2012)
73. Edson A (1962) Gas pipeline blast kills 8, 4 others hurt. The Montreal Gazette
74. NEWSGO.: Sombre anniversary in Lasalle: 50 years since gas explosion. CTV Montreal (2015)
75. Eldon, H.: Pipeline Explosion Kills Two Men. Sundre, Canada, Ottawa Citizen (1965)
76. Gordon, D.M.: Pipeline Explosion In Mexico Kills 52. The Telegraph, November 3 (1978)
77. Press, T.A.: 26 Killed in Mexico pipeline fire. CBCNEWS WORLD (2012)
78. Mosco, A.: Careless workers blamed for explosion. Observer-Reporter, June 6 (1989)
79. Mexico Pipeline Explosion Kills 26 Near US Border. Fox News (2012)
80. Anya, L.: Officials: W.Va. explosion was along newly installed natural gas line. Pittsburgh Post-Gazette Archived (2019)
81. Chris, L.: Explosion on Marshall County gas line heard and seen for miles.
82. Anya, L.: Landslide Caused West Virginia Pipeline Explosion. TransCanada reports, Pittsburgh Post-Gazette Archived (2018)
83. Jade, W.: Pipeline Explodes In Oklahoma. Huffington Post (2013)
84. Paul, S.: Nat Gas Pipeline Explosion Reported In Oklahoma (2016).
85. Millage, K.: Timeline of Bellingham pipeline explosion. Bellingham Herald (2016)
86. Ayodele, D.: Military Launches Attack on Militants Base in Ogun Coastal Communities. Information Nigeria (2017)
87. Stevenson, M.: Mexico Blast a Blow to Pemex's Improving Safety (2012)
88. Kinder Morgan must defend fatal Mexican gas pipeline explosion court case (2016)
89. Langford, C.: Kinder Morgan on Hook for Blast That Killed 22. Courthouse News Service (2016)
90. Ellingwood, K.: 27 Die in Oil Pipeline Explosion in Mexico. The Los Angeles Times Retrieved December 22 (2010)
91. Mexican pipeline explosion all but wipes out tiny town UPI. Morning News (Wilmington, Delaware: 2 (1978)
92. Government of Canada, Transportation Safety Board of Canada
93. https://www.reuters.com/article/us-pemex-fire/fire-at-mexico-pemex-gas-facility-kills-26-idUSBRE88H1DA20120919 (2012)
94. No Alarm Sounded When The West Virginia Pipeline Exploded : The Two-Way. NPR 2012-12-13 Retrieved 2016-10-04
95. Va, W.: gas explosion burns homes, shuts I-77. usatoday
96. Clayton, M.: West Virginia Gas Pipeline Explosion-Just a Drop in the Disaster Bucket. The Christain Science Monitor (2012)
97. San Bruno Explosion: Photos Of The Fire's Aftermath Paint A Bigger Picture. Retrieved on 8 Nov. 2011
98. Pipeline Accident Report, Natural Gas-Fuelled Building Explosion and Resulting Fire. New York City, National Transportation Safety Board, NTSB/PAR-15/01 PB2015–104889
99. Sanchez, R.: New York Explosion Exposes Nation's Aging and Dangerous Gas Mains. CNN (2014)
100. Groeger, L.: Pipelines Explained: How Safe are America's 2.5 Million Miles of Pipelines ProPublica (2012)
101. Watts, J.: China's Worst-Ever Oil Spill Threatens Wildlife as Volunteers Assist in Clean-Up. The Guardian (2010)
102. Chen, C.: Huge Pipeline Explosion in Northeastern China Causes Oil Spill. EPOCH TIMES (2015)
103. Edukugho, E.: Oil Pipeline Vandalism: What We Lost. Vanguard (2013)
104. Usman, E.: How vandals ambush, kill five JTF operatives in Arepo. Vanguard (2016)

105. Canadian, P.: Canadian Natural Resources Pipeline Leaks Near Slave Lake. CBC News, The Canadian Press (2014)
106. Gemmell.: Alta Oil Pipeline Leaked 28,000 Barrels. The Canadian Press CBC News (2011)
107. Torres, N.: Murphy Oil Reports Alberta Condensate Leak Now 17,000. Petro Global News (2015)
108. Madrazo, C.: 12 Lose Lives in Vast Lake of Blazing Oil. The Sydney Morning Gerald, New York (1959)
109. Lannon, A.: Eight Settle Claims Over Kanawha County Gas Pipeline Explosion. News Work TV, TriStateupdate (2013)
110. Shea, P.: Nat Gas Pipeline Explosion Reported In Oklahoma. The ValueWalk 20 (2013)
111. Emmanuel, A.: Preventable Errors Led to Pipeline Spill, Inquire Finds. Michigan, The New York Times (2012)
112. Millage, K.: Timeline of Bellingham Pipeline Explosion. The Bellingham Herald (2009)
113. Foster, J.M.: Natural Gas Pipeline Explosion Levels Homes in Kentucky Town. CLIMATEPROGRESS (2014)
114. Bouchard, C., Harquail, M., Simpson, C., Tadros, W.A.: Natural Gas Pipeline Rupture,Trans Canada Pipeline Limited, . Transportation Safety Board of Canada, Report No;P96H0012 (1996)
115. China, D.: Chinese Oil Pipeline Explosion. Reuters, November 23 (2013)
116. Kadir, A.: Transmission Gas Pipeline Lecture Note, Gas Engineering and Management, School of Computing. University of Salford, Manchester, Science and Engineering (2005)
117. Beavers, A.J., Neil, G., Thompson, C.: Technologies External Corrosion of Oil and Natural Gas Pipelines, ASM Handbook, Vol.13C, Corrosion: Environments and Industries ASM International (2006)
118. C M Engineers' Data Book. Institution of Mechanical Engineers, 4th edn.
119. Hatton, G.J., Pulici, M., Curti, G., Mansueto, M.: Deepwater Natural Gas Pipeline Hydrate Blockage caused by a Seawater Leak. Offshore Technology Conference (2002)
120. Report on The Crack Caused by Ice Formation in the Velero Mckee Refinery by The United States Chemical Safety and Hazard Investigation Board (CSB) (2008)
121. Fire From Ice. System failure case study **3** (08) (2009)
122. PRCI Pipeline Leak Detection Operation Improvements an Overview of Currently Available Leak Detection Technologies and US Regulations, Standards. In: 2011 Paper presented at the 6th Pipeline Technology Conference Hannover Germany
123. EQUIFAX.: Non Destructive Testing-Magnetic Particle Inspection. Werner Solken (2008–2016)
124. Beller, M.: Pipeline Inspection Utilizing Ultrasound Technology: On the Issue ofResolution. Pigging Products and Services Association (2007)
125. ESAB Knowledge Center, Radiographic and Ultrasonic Testing of Welds. Welding Inspection (2014)
126. Chis, T.: Pipeline Leak Detection Techniques. Anale Seria Informatica V (2007)
127. Davidson, R.: An introduction to Pipeline PIGGING. Halliburton Pipeline and Process Services, PIGing Products and Services Association (2002)
128. Geiger, G., Werner, T., Matko, D.: Leak detection and locating-a survey. In: PSIG annual meeting, 2003. Pipeline Simulation Interest Group (2003)
129. Kiss, K., Ranganath, S.: Online monitoring to assure structural integrity of nuclear reactor components. Int. J. Press. Vessels Pip. **34**(1–5), 3–15 (1988)
130. Ortiz-Villafuerte, J., Castillo-Duran, R., Alanso, G., Calleros-Micheland, G.: online monitoring system based on noise analysis. Nucl. Eng. Des. **236**(220), 2394–2404 (2006)
131. PHMSA Pipeline and Hazardous Materials Safety Administration—Guidance Manual for Operators of Small Natural Gas Systems Method of Detection A Leak, 4 (2002)
132. Razvarz, S., Jafari, R., Vargas-Jarillo, C.: Modelling and analysis of flow rate and pressure head in pipelines. In: 2019 16th International Conference on Electrical Engineering, Computing Science and Automatic Control (CCE), 11–13 Sept. 2019, pp. 1–6 (2019)

133. Jafari, R., Razvarz, S., Vargas-Jarillo, C., Yu, W.: Control of flow rate in pipeline using PID controller. In: 2019 IEEE 16th International Conference on Networking, Sensing and Control (ICNSC), 9–11 May 2019, pp 293–298 (2019)
134. Razvarz, S., Vargas-Jarillo, C., Jafari, R., Gegov, A.: Flow control of fluid in pipelines using PID controller. IEEE Access **7**, 25673–25680 (2019)
135. Razvarz, S., Vargas-Jarillo, C., Jafari, R.: Pipeline monitoring architecture based on observability and controllability analysis. In: 2019 IEEE International Conference on Mechatronics (ICM), 18–20 March 2019, pp 420–423 (2019)
136. Razvarz, S., Jafari, R., Vargas-Jarillo, C., Gegov, A., Forooshani, M.: Leakage detection in pipeline based on second order extended Kalman filter observer. IFAC-PapersOnLine **52**(29), 116–121 (2019). https://doi.org/10.1016/j.ifacol.2019.12.631
137. Papadopoulou, K.A., Shamount, M.N., Lennox, B., Mackay, D., Turner, J.T., Wang, X.: An evaluation of acoustic reflectometry for leakage and blockage detection. IMechE (2008). https://doi.org/10.1243/09544062JMES873
138. Newell, R.D.: Mass Balance Leak Detect. Can it Work for You, ENTELEC (2006)
139. Geiger, G., Werner, T., Matko, P.: Leak detection and locating-a survey. In: 35th Annual PSIG Meeting, Bern, Switzerland (2003)
140. Corp, A.: New Leak Detection and Monitoring Technology Ensures of Pipelines. Press Release (2010)
141. Billmann, L., Isermann, R.: Leak detection methods for pipelines. IFAC Proc. **17**(2), 1813–1818 (1984)
142. Siebert, H., Klaiber, T.: Testing a method for leakage monitoring of a gasoline pipeline. Process Autom. **5**, 91–96 (1980)
143. Verde, C.: Minimal order nonlinear observer for leak detection. J. Dyn. Sys. Mcas. Control **126**(3), 467–472 (2004)
144. Verde, C.: Multi-leak detection and isolation in fluid pipelines. Control Eng. Pract. **9**(6), 673–682 (2001)
145. Verde, C., Sánchez-Parra, M.: Application of structural analysis to improve fault diagnosis in a gas turbine. Gas Turbines: 307 (2010)
146. Verde, C.: Leakage location in pipelines by minimal order nonlinear observer. In: Proceedings of the 2001 American Control Conference.(Cat. No. 01CH37148), pp. 1733–1738. IEEE (2001)
147. Carrera, R., Verde, C., Cayetano, R.: A SCADA expansion for leak detection in a pipeline. Sensors **2300**(2320), 2340 (2015)
148. Angulo, M.T., Verde, C.: Second-order sliding mode algorithms for the reconstruction of leaks. In: 2013 Conference on Control and Fault-Tolerant Systems (SysTol), pp. 566–571. IEEE (2013)
149. Aamo, O.M., Smyshlyaev, A., Krstic, M., Foss, B.A.: Output feedback boundary control of a Ginzburg-Landau model of vortex shedding. IEEE Trans. Autom. Control **52**(4), 742–748 (2007)
150. Hauge, E,. Aamo, O.M., Godhavn, J.-M.: Model based pipeline monitoring with leak detection. In: Seventh IFAC Symposium on Nonlinear Control Systems, 2007. vol 1 (2007)
151. Jolly, W.D., Morrow, T.B., O'Brien, J.F., Spence, H.F., Svedeman, S.J.: New Methods for Rapid Leak Detection in Offshore Pipelines Reports Prepared for the Minerals Management Service US Department of Interior (1992)
152. Anastasopoulos, A., Kourousis, D., Bollas, K.: Acoustic emission leak detection of liquid filled buried pipeline. J. Acoust. Emiss. **27** (2009)
153. Fuchs, H.V., Riehle, R.: Ten years of experience with leak detection by acoustic signal analysis. Appl. Acoust. **33**(1), 1–19 (1991). https://doi.org/10.1016/0003-682X(91)90062-J
154. Randall, R.B.: Frequency Analysis. Bruel and Kjaer (1987)
155. Tashvaei, M., Beck, S.B., Staszewski, A.: Leakage detection in pipelines using cepstrum analysis. Inst. Phys. Publ. Meas Sci. Technol. **17**, 367–372 (2006)
156. Beck, S.B.M., Foong, J., Staszewski, A.: Wavelet and Ceptrum analyses of leaks in pipe networks. Progr. Indus. Math. ECMI (2004)

157. Santos, R.B., Almeida, W.S., Silva, FVd., Cruz, SLd., Fileti, A.M.F.: Spectral analysis for detection of leaks in pipes carrying compressed air. Chem. Eng. Trans. **32**, 1363–1368 (2013). https://doi.org/10.3303/CET1332228
158. Sarkar, DaS.: Review of pipeline hazard detection using vibration analysis method. Int. J. Mech. Prod. Eng. **3**(1) (2015)
159. Lay-Ekuakille, A., Vergallo, P., Trotta, A.: Impedance method for urban water works: experimental frequency analysis for leakage detection. 15th IWADC Workshop (2010)
160. Lotfollahi-Yaghin, M.A., Hesari, M.A.: Using wavelet analysis in crack detection at the arch concrete dam under frequency analysis with FEM. World Appl. Sci. J. **3**(4), 691–704 (2008)
161. Liew, K.M., Wang, Q.: Application of wavelet theory for crack identification in structure. J. Eng. Mech. **124**, 152–157 (1998a)
162. Kim, H., Melhem, H.: Damage detection of structures by wavelet analysis Eng. Struct. **26**, 347–362 (2004)
163. Zhong, S., Oyadiji, S.O.: Detection of cracks in S imply-supported beams by continuous wavelet transform of reconstructed modal data. Comput. Struct. **89**, 127–148 (2011)
164. Gentile, A., Messina, A.: On the continuous wavelet transforms applied to discrete vibrational data for detecting open cracks in damaged beams. Int. J. Solids Struct. **40**, 295–315 (2003)
165. Rucka, M., Wilde, A.: Application of continuous wavelet transform in vibration based damage detection method for beams and plates. J. Sound Vib. **297**, 536–550 (2006)
166. Hong, J.C., Kim, Y.Y., Lee, H.C., Lee, Y.W.: Damage detection using the Lipschitz exponent estimated by the wavelet transform; applications to vibration modes of a beam. Int. J. Solids Struct. **39**, 1803–1816 (2002)
167. Kulkarni, P.G., Sahasrabudhe, A.D.: Application of wavelet transform for fault diagnosis of rolling element bearings. Int. J. Sci. Technol. Res. **2**(4) (2013)
168. Zhong, S., Oyadiji, S.O.: Crack detection in simply supported beams using stationary wavelet transform of modal data. J. Vib. Acoust. **18**, 169–190 (2011a)
169. Zhong, S., Oyadiji, S.O.: Crack detection in simply supported beams without baseline modal parameters by stationary wavelet transform of modal data. Mech. Syst. Sig. Process. **21**, 1853–1884 (2007)
170. Fan, X., Zuo, M.J., Wang, X.: Identification of weak ultrasonic signals in testing of metallic materials using wavelet transform. Inst. Phys. Publ. Smart Mater Struct. **15**, 1531–1539 (2006)
171. Ma, M., Liang, J., Sun, L., Wang, M.: SAR image segmentation based on SWT and improved AFSA. Third International Symposium on Intelligent Information Technology and Security Informatics (2010). https://doi.org/978-0-7695-4020-7/10
172. Prats-Montalbána, J.M., Cocchib, M., Ferrer, A.: N-way modelling for wavelet filter determination in multivariate image analysis. J. Chemo Metr. **29**, 379–388 (2015)
173. Seker, S., Karatoprak, E., Kayran, A.H., Senguler, T.: Stationary wavelet transform for fault detection in rotating machinery. Proc of SPIE 6763 (2007)
174. Jamaluddin, F.N., Ahmad, S.A., Bahari, S., Noor, M., Hassan, W.Z.W., Yaacob, A., Adam, Y.: Performance of DWT and SWT in muscle fatigue detection. In: 2015 IEEE Student Symposium in Biomedical Engineering and Science (ISSBES) (2015)
175. Backus, J.: The Acoustic Foundations of Music, 2nd edn. WW Norton and Company, New York (1977)
176. Mortazavi, S.H., Shahrtash, S.M.: Comparing de-noising performance of DWT, WPT, SWT and DT-CWT for partial discharge signal. J. Univers. Power Eng. Conf., 1–6 (2008)
177. Zhong, S., Oyadiji, S.O.: Crack detection in simply supported beams using stationary wavelet transform of modal data. ASME J. Vibr. Acoust. **18**, 169–190 (2009)
178. Zhong, S., Oyadiji, S.O.: Detection of cracks in simply-supported beams by continuous wavelet transform of reconstructed modal data. Comput. Struct. **89**, 127–148 (2011)
179. Mortazavi, S.H., Shahrtash, S.M.: Comparing de-noising performance of DWT,WPT, SWT and DT-CWT for partial discharge signal. J. Univers. Power Eng. Conf., 1–6 (2008)
180. Chen, Y., Ma, H.: Signal de-noising in ultrasonic testing based on stationary wavelet transform. WRI Global Congr. Intell. Syst. (2009). https://doi.org/10.1109/GCIS.2009.266

181. Zhong, S., Oyadiji, S.O.: Identification of cracks in beams with auxiliary mass spatial probing by stationary wavelet transform. ASME J. Vibr. Acoust. **130**, 1–14 (2008)
182. Seker, S., Karatoprak, E., Kayran, A.H., Senguler, T.: Stationary wavelet transform for fault detection in rotating machinery. Proc. SPIE **6763**, 67630A (2007)
183. Lotfollahi-Yaghin, M.A., Hesari, M.A.: Using wavelet analysis in crack detection at the arch concrete dam under frequency analysis with FEM. World Appl. Sci. J. **3**(4) (2008)
184. Liew, K.M., Wang, Q.: Application of wavelet theory for crack identification in structures. J. Eng. Mech. **124**, 152–157 (1998b)
185. Kim, H., Melhem, H.: Damage detection of structures by wavelet analysis. Eng. Struct. **26**, 347–362 (2004)
186. Zhong, S.O., Yadiji, S.O.: Detection of cracks in simply-supported beams by continuous wavelet transform of reconstructed modal data. Comput. Struct. **89**, 127–147 (2011)
187. Kulkarni, P.G., Sahasrabudhe, A.D.: Application of wavelet transform for fault diagnosis of rolling element bearings. Int. J. Sci. Technol. Res. **2**(4) (2013)
188. Katunin, A.: Non-destructive damage assessment of composite structures based on wavelet analysis of modal curvatures: State-of-the-Art review and description of wavelet-based damage assessment benchmark, pp. 1–19. Hindawi Publishing Corporation Shock and Vibration (2015)
189. Zhong, S., Oyadiji, S.O.: Crack detection in simply supported beams using stationary wavelet transform of modal data. ASME J. Vibr. Acoust. **18**, 169–190 (2011b)
190. Nagendra, H., Mukherjee, S., Kumar, V.: Application of wavelet techniques in ECG signal processing: an overview. Int. J. Eng. Sci. Technol. (IJEST) **3** 10) (2011)
191. Kshirsagar, P., Salodkar, A., Bhaiswar, R.: Iris recognition using stationary wavelet transform and artificial neural network. Int. J. Eng. Innov. Technol. **1**(3) (2012)
192. Zhong, S., Oyadiji, S.O.: Analytical predictions of natural frequencies of crack simply supported beams with a stationary roving mass. J. Sound Vib. **311**, 328–352 (2008)
193. Wang, Y., He, Z., Zi, Y.: A demodulation method based on improved local mean decomposition and its application in rub-impact fault diagnosis. Meas. Sci. Technol. **20**, 1–10 (2009). https://doi.org/10.1088/0957-0233/20/2/02570
194. Sakamoto, J.M.S., Baba, A., Tittmann, B.R., Mulry, J., Kropf, M., Pacheco, G.M.: Non-destructive inspection of a composite material sample using a laser ultrasonic system with a beam homogenizer. Rev. Progr. Quant. Nondestruct. Eval. **30**(1335), 935–941 (2011). https://doi.org/10.1063/1.3592038

Chapter 3
Modelling of Pipeline Flow

3.1 Introduction

As the needs of pipeline industries evolved over the years, the large lines that move gas and liquid long distances made. Pipeline systems are among the safest and biggest transportation services, but that does not indicate that they are risk-free. Environmental threats might lead to damage, leakage, or blockage in pipe systems that consequently cause significant economic losses. Thus, effective quality assurance is necessary for pipeline companies.

In [1] a leak detection system consisting of an adaptive observer is developed based on pressure and flow measurements at the entry and the exit of the pipe. In [2–8] a bank of observers is designed using the algorithm developed in [9]. In [10] it has been proved that because of the uncomplicated construction of the discretized model the observer can, therefore, be designed plainly. Therefore, the requirement for recalculating the bank of observers could be ignored if the functioning point of the piping system alters. In [11], two nonlinear minimal order observers are utilised to obtain the residual with uncertainty, and also the leakage location is estimated from a steady-state relationship between residual, leakage flow, and uncertainty. The leak identification technique proposed in [12–14] is capable of identifying multiple leakages utilising a bank of observers. In [15–17] an exponentially convergent observer is developed for a linearized Ginzburg–Landau structure of vortex shedding in the bluff body flows where the phenomenon of vortex shedding occurs. Today, a variety of methods and techniques such as neural networks, deep learning, fuzzy logic, etc. are available with a vast range of applications in engineering [18–40]. These techniques have been successfully used for leak detection. The most common fault detection approaches include observer-based and identification-based schemes [41–48].

In this chapter, a simple pipeline is analysed with no changes in temperature and density. The entrance and exit flow rate and pressure are assumed to be measurable variables. Furthermore, the convective changes in fluid velocity in a pipe are neglected, and the cross-section area of the pipe as well as the fluid density are

S. Razvarz et al., *Flow Modelling and Control in Pipeline Systems*, Studies in Systems, Decision and Control 321, https://doi.org/10.1007/978-3-030-59246-2_3

supposed to be constant. The fluid motion in the pipe is described by continuity and momentum partial differential equations.

3.2 Lagrangian and Eulerian Specification of the Flow Field

Classical field theory describes the study of how one or more physical fields interact with matter through field equations.

3.2.1 Lagrangian Field

Lagrangian view of the flow field is a way of looking at fluid motion in which the observer follows an individual fluid parcel as it moves [49, 50]. The individual fluid particles are marked, and their positions, velocities, etc. are defined as a function of time. The Lagrangian fluid parcel can be represented by some time-independent vector field x_0. Based on the Lagrangian description, the flow can be defined by a function $X(x_0, t)$ where the position vector x_0 is shown at a given time t.

3.2.2 Eulerian Field

In the Eulerian scheme, the motion of fluid particles at a given point in space is defined [49, 50]. The velocity in the fluid,$v(x, t)$ is a function of both position x and time t.

3.2.2.1 The Euler Equation

The formula to determine the force on any fluid volume is defined as: $\mathcal{F} = \int \wp ds$ $\left[\mathrm{N} = \mathrm{N/m^2} * \mathrm{m^2}\right]$ or $\left[\mathrm{kg\,m/s^2} = \mathrm{kg/m\,s^2} * \mathrm{m^2}\right]$. Therefore the force on a unit volume is defined as [50]:

$$\mathcal{F} = \rho \frac{dv}{dt} \tag{3.1}$$

where ρ is the density of fluid. Using the chain rule the following is obtained:

$$dv = dt\frac{\partial v}{\partial t} + dr.\nabla v = dt\frac{\partial v}{\partial t} + dx\frac{\partial v}{\partial x} + dy\frac{\partial v}{\partial y} + dz\frac{\partial v}{\partial z} \tag{3.2}$$

in which $dr = udt$ is the distance moved by the fluid particle during dt. Hence,

$$dv = dt\left[\frac{\partial v}{\partial t} + \frac{dx}{dt}\frac{\partial v}{\partial x} + \frac{dy}{dt}\frac{\partial v}{\partial y} + \frac{dz}{dt}\frac{\partial v}{\partial z}\right] \tag{3.3}$$

Dividing (2) by dt the following is extracted:

$$\frac{dv}{dt} = \frac{\partial v}{\partial t} + v.\nabla v \tag{3.4}$$

3.2.3 *Modeling of Liquid Flow in the Pipeline*

3.2.4 *Momentum Equation*

Writing Newton's second law for a unit mass of a fluid, the following Euler's equation is obtained

$$\mathcal{F} = ma \tag{3.5}$$

we have $\mathcal{F} = \nabla\wp$ such that $\mathcal{F}$ is taken to be the force density and is the gradient of pressure. Force density is defined as follows:

$$\nabla\wp = \rho\frac{dv}{dt} = \rho\left(\frac{\partial v}{\partial t} + v.\nabla v\right) \tag{3.6}$$

$$\frac{\partial v}{\partial t} + v.\nabla v = \frac{\nabla\wp}{\rho} \tag{3.7}$$

where P is the pressure. An external body force can be added to (3.7) as follows:

$$\frac{\partial v}{\partial t} + v.\nabla v = \frac{\nabla\wp}{\rho} + \mathcal{F}_b \tag{3.8}$$

We have,

$$v.\nabla v = 0 \tag{3.9}$$

Based on the definition of dot product [51, 52],

$$a.b = ab\cos\theta \tag{3.10}$$

we have $u.\nabla u = 0$. The body force can be defined as [53]:

$$s = \frac{\daleth}{2\eth} v^2 \tag{3.11}$$

where $\eth$ is the diameter (m) of the pipe and also $\daleth$ is the friction coefficient in the pipe. Hence,

$$\frac{\partial v}{\partial t} + \frac{1}{\rho}\frac{\partial \wp}{\partial z} + \frac{\daleth}{2\eth} v^2 = 0 \tag{3.12}$$

Thus,

$$\frac{\partial v}{\partial t} + \frac{1}{\rho}\frac{g}{g}\frac{\partial \wp}{\partial z} + \frac{\daleth}{2\eth} v^2 = 0 \tag{3.13}$$

We have,

$$\mathring{A}\left[\frac{\partial v}{\partial t} + \frac{1}{\rho}\frac{g}{g}\frac{\partial \wp}{\partial z} + \frac{\daleth}{2\eth}\frac{\mathring{A}^2}{\mathring{A}^2} v^2 = 0\right] \tag{3.14}$$

where $\mathring{A}$ is the section area $\left(m^2\right)$. Therefore,

$$\mathring{A}\frac{\partial v}{\partial t} + \mathring{A}\frac{g}{\rho g}\frac{\partial \wp}{\partial z} + \frac{\mathring{A}\daleth}{2\eth}\frac{\mathring{A}^2}{\mathring{A}^2} v^2 = 0 \tag{3.15}$$

Hence,

$$\frac{\partial \mathring{A} v}{\partial t} + \mathring{A} g\frac{\partial}{\partial z}\frac{\wp}{\rho g} + \frac{\mathring{A}\daleth}{2\eth}\frac{\mathring{A}^2}{\mathring{A}^2} v^2 = 0 \tag{3.16}$$

Let fluid flow and pressure head are $\mathbb{Q} = \mathring{A} v$ and $\mathcal{H} = \frac{\wp}{\rho g}$ respectively, hence Eq. (3.16) can be rewritten as follows:

$$\frac{\partial \mathbb{Q}}{\partial t} + \mathring{A} g\frac{\partial}{\partial z}\mathcal{H} + \frac{\daleth}{2\eth}\frac{\mathbb{Q}^2}{\mathring{A}} = 0 \tag{3.17}$$

where z is the length coordinate (m), t is the time coordinate (s), and g is the gravity $\left(\mathrm{m/s^2}\right)$.

3.2.5 Continuity Equation

The continuity equation can be derived by applying the principle of conservation of mass to a control volume [54–58].

Suppose $\beth$ denotes the momentum of fluid and $\gimel$ denotes the corresponding intensive property. An intensive property can be described as the quantity of B per unit mass of a system, for example, $\gimel = \lim_{\Delta m \to 0} \frac{\Delta \beth}{\Delta m}$.. The total quantity of B in a control volume, B_{cv}, can be defined as:

$$\beth_{cv} = \int_{cv} \gimel \rho d\forall \tag{3.18}$$

where m is mass, ρ is mass density and $d\forall$ is the differential volume of the fluid. Property $\beth$ of the system at times t and $t + \Delta t$ can be defined as follows:

$$\begin{aligned} \beth_{sys}(t) &= \beth_{cv}(t) + \Delta\beth_{in} \\ \beth_{sys}(t + \Delta t) &= \beth_{cv}(t + \Delta t) + \Delta\beth_{out} \end{aligned} \tag{3.19}$$

In which the subscripts sys and cv stand for the system and the control volume respectively. For any control volume the subscripts in and out mean the inflow and outflow, respectively. Also, $\Delta\beth_{in}$ and $\Delta\beth_{out}$ stand for inflow and outflow of property $\beth$ going in and out of the volume, respectively. We have,

$$\frac{d\beth_{sys}}{dt} = \lim_{\Delta t \to 0} \frac{\beth_{sys}(t + \Delta t) - \beth_{sys}(t)}{\Delta t} \tag{3.20}$$

By replacing (3.19) into (3.20) we have,

$$\frac{d\beth_{sys}}{dt} = \lim_{\Delta t \to 0} \frac{\beth_{cv}(t + \Delta t) - \beth_{cv}(t)}{\Delta t} + \lim_{\Delta t \to 0} \frac{\beth_{out}(t)}{\Delta t} - \lim_{\Delta t \to 0} \frac{\beth_{in}(t)}{\Delta t} \tag{3.21}$$

Since Δt goes towards zero, the first term on the right-hand side of (21) demonstrates the time rate of change of B within the control volume, for example,

$$\frac{d\beth_{cv}}{dt} = \lim_{\Delta t \to 0} \frac{\beth_{cv}(t + \Delta t) - \beth_{cv}(t)}{\Delta t} \tag{3.22}$$

By replacing (3.18) into (3.22) we have,

$$\lim_{\Delta t \to 0} \frac{\beth_{cv}(t + \Delta t) - \beth_{cv}(t)}{\Delta t} = \frac{d}{dt} \int_{cv} \gimel \rho d\forall \tag{3.23}$$

We have,

$$\begin{aligned} \lim_{\Delta t \to 0} \frac{\Delta v\, \beth_{out}(t)}{\Delta t} &= \left(\gimel \rho \mathring{A} V_s\right)_{out} \\ \lim_{\Delta t \to 0} \frac{\Delta v\, \beth_{in}(t)}{\Delta t} &= \left(\gimel \rho \mathring{A} V_s\right)_{in} \end{aligned} \tag{3.24}$$

such that Å represents the cross-sectional area of the conduit and V_s represents the flow velocity measured with respect to the control surface.

$$\frac{d\beth_{sys}}{dt} = \frac{d}{dt}\int_{cv} \lambda\rho d\forall + \left(\lambda\rho\mathring{A}V_s\right)_{out} - \left(\lambda\rho\mathring{A}V_s\right)_{in} \tag{3.25}$$

If the control volume is constant, then V_s represents the fluid flow velocity. Nevertheless, in a case that the control volume is contracted relative to time, then V_s in (3.25) is considered as the relative flow velocity and can be defined as $V_s = (V - W)$, where W represents the velocity of the control surface across the inlet and outlet. V as well as W can be calculated with regard to the coordinate axes. Therefore, (3.25) for a contracting control volume in a one-dimensional flow can be rewritten as follows:

$$\frac{d\beth_{sys}}{dt} = \frac{d}{dt}\int_{cv} \lambda\rho d\forall + (\lambda\rho A(V - W))_{out} - (\lambda\rho A(V - W))_{in} \tag{3.26}$$

By applying the Reynolds transport theorem for the conservation of mass, the intensive property of the fluid can be defined as $\lambda = \lim_{\Delta m \to 0} \frac{\Delta m}{\Delta m} = 1$.. Furthermore, the mass of the system is taken to be fixed, $\frac{dM_{sys}}{dt} = 0$. If we apply (3.26) to the control volume and supposing $\lambda = 0$, the following can be extracted,

$$\frac{d}{dt}\int_{cv} \lambda\rho d\forall + \rho_2\mathring{A}_2(V_2 - W_2) - \rho_1\mathring{A}_1(V_1 - W_1) = 0 \tag{3.27}$$

By applying the Leibniz's rule [59] to the first term on the left side the following relation can be obtained,

$$\frac{d}{dt}\int_{f_1(t)}^{f_2(t)} \frac{\partial}{\partial t}F(x,t) + F(f_2(t),t)\frac{df_2}{dt} - F(f_1(t),t)\frac{df_1}{dt} \tag{3.28}$$

Also,

$$\int_{x_1}^{x_2} \frac{\partial}{\partial t}(\rho\mathring{A})dx + \rho_2\mathring{A}_1\frac{dx_1}{dt} + \rho_2\mathring{A}_2(V_2 - W_2) - \rho_1\mathring{A}_1(V_1 - W_1) = 0 \tag{3.29}$$

Since $\frac{dx_2}{dt} = W_2$ and $\frac{dx_1}{dt} = W_1$, therefore,

$$\int_{x_1}^{x_2} \frac{\partial}{\partial t}(\rho \text{Å})dx + \rho_2 \text{Å}_2 V_2 - \rho_1 \text{Å}_1 V_1 = 0 \tag{3.30}$$

such that $\Delta x = x_2 - x_1$. If both sides of the above relation divide by Δx and Δx goes towards zero, the following relation can be obtained,

$$\frac{\partial \text{Å}}{\partial t}(\rho \text{Å}) + \frac{\partial}{\partial x}(\rho \text{Å} V) = 0 \tag{3.31}$$

Hence,

$$\text{Å}\frac{\partial \rho}{\partial t} + \rho\frac{\partial \text{Å}}{\partial t} + \rho \text{Å}\frac{\partial V}{\partial x} + \rho V\frac{\partial \text{Å}}{\partial x} + \text{Å} V\frac{\partial \rho}{\partial x} = 0 \tag{3.32}$$

Therfore,

$$\frac{1}{\rho}\frac{d\rho}{dt} + \frac{1}{\text{Å}}\frac{d\text{Å}}{dt} + \frac{\partial V}{\partial x} = 0 \tag{3.33}$$

We are interested in obtaining the pressure intensity p and the flow velocity V. We display the derivatives of ρ and Å with respect to $\wp$ and V. In a fluid, the bulk modulus can be defined as:

$$\kappa = \frac{d\wp}{d\rho/\rho} \tag{3.34}$$

Hence,

$$\frac{d\rho}{dt} = \frac{\rho}{\kappa}\frac{d\wp}{dt} \tag{3.35}$$

Also, for a circular conduit with radius r we have,

$$\frac{d\text{Å}}{dt} = 2\pi r\frac{dr}{dt} \tag{3.36}$$

The above relation can be rewritten as follows:

$$\frac{d\text{Å}}{dt} = 2\pi r^2\frac{1}{r}\frac{dr}{dt} \tag{3.37}$$

or

$$\frac{1}{\text{Å}}\frac{d\text{Å}}{dt} = 2\frac{d\epsilon}{dt} \tag{3.38}$$

Let us suppose that the conduit walls are linearly elastic, hence,

$$\epsilon = \frac{\sigma_2 - \mu\sigma_1}{E} \tag{3.39}$$

such that σ_2 is hoop stress [60], σ_1 is axial stress and μ is Poisson ratio. By assuming $\sigma_1 = 0$, Eq. (3.39) can be written as follows:

$$\epsilon = \frac{\sigma_2}{E} \tag{3.40}$$

Also we have,

$$\sigma_2 = \frac{\wp\eth}{2e} \tag{3.41}$$

in which $\wp$ is taken to be the inside pressure, e is taken to be the thickness of the conduit walls and $\eth$ is taken to be the conduit diameter. Equation (3.41) can be rewritten as follows:

$$\frac{d\sigma_2}{dt} = \frac{\wp}{2e}\frac{d\eth}{dt} + \frac{\eth}{2e}\frac{d\wp}{dt} \tag{3.42}$$

Hence,

$$E\frac{d\epsilon}{dt} = \frac{\wp}{2e}\frac{d\eth}{dt} + \frac{\eth}{2e}\frac{d\wp}{dt} \tag{3.43}$$

Therefore,

$$\frac{d\epsilon}{dt} = \frac{\frac{\eth}{2e}\frac{d\wp}{dt}}{E - \frac{\wp\eth}{2e}} \tag{3.44}$$

Applying Eqs. (3.40) and (3.44) the following is obtained,

$$\frac{1}{\AA}\frac{d\AA}{dt} = \frac{\frac{\eth}{2e}\frac{d\wp}{dt}}{E - \frac{\wp D}{2e}} \tag{3.45}$$

Replacing (3.35) and (3.45) into (3.33), we have,

$$\frac{\partial V}{\partial x} + \left(\frac{1}{\kappa} + \frac{1}{\frac{eE}{D} + \frac{\wp}{2}}\right)\frac{d\wp}{dt} = 0 \tag{3.46}$$

As $\wp/2 \ll eE/\eth$, hence,

$$\frac{\partial V}{\partial x} + \left(\frac{1}{\kappa} + \frac{1}{\frac{eE}{\eth\kappa}}\right)\frac{d\wp}{dt} = 0 \tag{3.47}$$

A pressure wave travels along the pipe at velocity a. We have,

$$a^2 = \frac{\frac{K}{\rho}}{1 + \frac{\eth K}{eE}} \tag{3.48}$$

Hence,

$$\frac{\partial\wp}{\partial t} + V\frac{\partial\wp}{\partial x} + \rho a^2\frac{\partial V}{\partial x} = 0 \tag{3.49}$$

Generally, $V(\partial V/\partial x)$ and $V(\partial\wp/\partial x)$ have small amounts and can be neglected. Thus,

$$\frac{\partial\wp}{\partial t} + \rho a^2\frac{\partial V}{\partial x} = 0 \tag{3.50}$$

In hydraulic engineering the discharge, $\mathbb{Q}$, is calculated as $\mathbb{Q} = V\text{Å}$ and the pressure intensity can be defined as follows:

$$\wp = \rho g(\mathcal{H} - z) \tag{3.51}$$

where z is the height above the datum and $\mathcal{H}$ is the pressure heads. For a horizontal pipe, we have $\frac{dz}{dx} = 0$. Therefore, following (51), we have $\partial\wp/\partial t = \rho g(\partial\mathcal{H}/\partial t)$ and $\partial\wp/\partial x = \rho g(\partial\mathcal{H}/\partial x)$. By replacing these relationships into (3.51), the following can be extracted,

$$\frac{\partial\mathcal{H}}{\partial t} + \frac{a^2}{g\text{Å}}\frac{\partial\mathbb{Q}}{\partial x} = 0 \tag{3.52}$$

Eventually, the continuity equation is demonstrated as follows:

$$\frac{\partial\mathcal{H}}{\partial t} + \frac{a^2}{g\text{Å}}\frac{\partial\mathbb{Q}}{\partial x} = 0 \tag{3.53}$$

in which a is taken to be the velocity of the pressure wave.

The pressure heads ($\mathcal{H}$) as well as flow rate ($\mathbb{Q}$) are functions of position and time, $\mathcal{H}(x, t)$ and $\mathbb{Q}(x, t)$, such that $x \in [0, L]$ and L is taken to be the length of pipe. Pressure heads at the beginning and end points of pipe are controllable and measurable initial and boundary conditions and are defined as:

$$\begin{cases} \mathcal{H}(0,t) = \mathcal{H}_{in}(t) \\ \mathcal{H}(L,t) = \mathcal{H}_{out}(t) \end{cases} \tag{3.54}$$

When the variation of flow rate is small enough, (3.17) can be rewritten as follows:

$$\frac{\partial \mathbb{Q}}{\partial t} + \mathring{A} g \frac{\partial}{\partial x}\mathcal{H} + \frac{Ⅎ\mathbb{Q}}{ð\mathring{A}} = 0 \tag{3.55}$$

3.3 Modeling of Flow in Pipeline

Leaks are typically detected and isolated using direct on-line observer-based approaches. Initially, the nonlinear model is designed and then some suitable input variations are selected for ensuring leakage recognisability. A common procedure to develop a suitable model for observer design is to utilise some finite-dimensional estimation, for example using finite differences for space discretization by partitioning the pipe into a supplied number of segments and assuming time as a continuous variable. Finally, the flow dynamics can be illustrated through the flow rates in each of the segments along the pipeline and the pressures at the end of every segment. The pipe model can be illustrated using (3.53) and (3.55).

Obtaining the solution of (3.53) and (3.55) are difficult. Nevertheless, various numerical techniques have been utilised to obtain the solutions of these equations [61, 62]. In this chapter, for modeling a pipeline we use finite difference technique. This technique is a simple procedure to design a more suitable nonlinear model as an observer and controllable require a good model of the system. The finite difference technique is the oldest among the discretization methods that partitions the pipe into N number of segments and assumes time as a continuous variable. The flow dynamics are illustrated through the flow rates in each of the segments along the pipeline and the pressures at the end of every segment. In Fig. 3.1, the pipe is divided into four segments such that the variables $\mathcal{H}_1, \mathcal{H}_2, \mathcal{H}_3, \mathcal{H}_4, \mathcal{H}_5$ are pressure heads

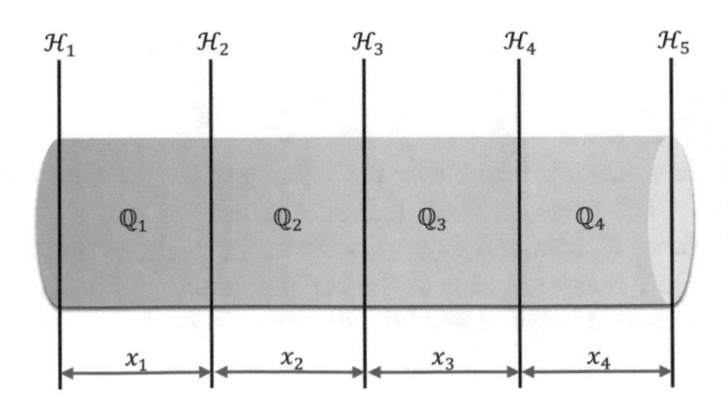

Fig. 3.1 Finite difference for a pipe having four segments and actuation at the start and end of each segment

and $\mathbb{Q}_1, \mathbb{Q}_2, \mathbb{Q}_3, \mathbb{Q}_4$ are flow rates. We can measure pressure heads ($\mathcal{H}_1$ *and* $\mathcal{H}_5$) and flow rates ($\mathbb{Q}_1$ *and* $\mathbb{Q}_4$) at the beginning and end of the pipe.

3.4 Steady State Model

3.4.1 Case 1

In this section, we aim to design a model for (3.54). A general linear model can be defined as[63–65]:

$$\frac{dx}{dt} = Ax + Bu \tag{3.56}$$

$$y = Cx \tag{3.57}$$

in which $x(t)$ is taken to be the state vector, $u(t)$ is the input and $y(t)$ is the output of the system. In our designed model we have supposed that $x = (\mathbb{Q}_1\mathbb{Q}_2\mathbb{Q}_3\mathbb{Q}_4\mathcal{H}_2\mathcal{H}_3\mathcal{H}_4)^T$,$u[\mathcal{H}_{in}, \mathcal{H}_{out}]^T = (\mathcal{H}_1\mathcal{H}_5)^T$ and $y = (\mathbb{Q}_1\mathbb{Q}_4)^T$. We aim to obtain $\dot{x}$ from (53). Therefore,

$$\frac{\partial\mathcal{H}(x,t)}{\partial t} + \frac{a^2}{g\mathring{A}}\frac{\partial\mathbb{Q}(x,t)}{\partial x} = 0 \tag{3.58}$$

Hence,

$$\frac{\partial\mathcal{H}(x,t)}{\partial t} = -\frac{a^2}{g\mathring{A}}\frac{\partial\mathbb{Q}(x,t)}{\partial x} \tag{3.59}$$

The above relation can be rewritten as follows:

$$\frac{\partial\mathcal{H}(x,t)}{\partial t} = -\frac{a^2}{g\mathring{A}\Delta x}\Delta\mathbb{Q}(x,t) \tag{3.60}$$

Thus,

$$\frac{\partial\mathcal{H}(x_i,t)}{\partial t} = -\frac{a^2}{g\mathring{A}\Delta x}(\mathbb{Q}(x_i,t) - \mathbb{Q}(x_{i-1},t)) \tag{3.61}$$

We can rewrite (3.55) as follows:

$$\frac{\partial\mathbb{Q}}{\partial t} = -\mathring{A}g\frac{\partial\mathcal{H}}{\partial x} - \frac{\text{Ⅎ}\mathbb{Q}}{\eth\mathring{A}} \tag{3.62}$$

Hence,

$$\frac{\partial \mathbb{Q}}{\partial t} = \frac{-\text{Å}g}{\partial x} \partial \mathcal{H} - \frac{\Finv \mathbb{Q}}{\eth \text{Å}} \tag{3.63}$$

Applying the difference between two points, the above relation can be rewritten as:

$$\frac{\partial \mathbb{Q}}{\partial t} = \frac{-\text{Å}g}{\Delta x} \Delta \mathcal{H} - \frac{\Finv \mathbb{Q}}{\eth \text{Å}} \tag{3.64}$$

Thus,

$$\frac{\partial \mathbb{Q}(x_i, t)}{\partial t} = \frac{-\text{Å}g}{\Delta x} (\mathcal{H}(x_i, t) - \mathcal{H}(x_{i-1}, t)) - \frac{\Finv}{\eth \text{Å}} \mathbb{Q}(x_i, t) \tag{3.65}$$

Eventually, from Eqs. (3.61) and (3.65) we have,

$$\begin{array}{ll} \dot{\mathcal{H}}_1 = \frac{-a^2}{g\text{Å}\Delta x}(\mathbb{Q}_1 - \mathbb{Q}_0) & \dot{\mathbb{Q}}_1 = \frac{-\text{Å}g}{\Delta x}(\mathcal{H}_2 - \mathcal{H}_1) - \frac{\Finv}{\eth \text{Å}} \mathbb{Q}_1 \\ \dot{\mathcal{H}}_2 = \frac{-a^2}{g\text{Å}\Delta x}(\mathbb{Q}_2 - \mathbb{Q}_1) & \dot{\mathbb{Q}}_2 = \frac{-\text{Å}g}{\Delta x}(\mathcal{H}_3 - \mathcal{H}_2) - \frac{\Finv}{\eth \text{Å}} \mathbb{Q}_2 \\ \dot{\mathcal{H}}_3 = \frac{-a^2}{g\text{Å}\Delta x}(\mathbb{Q}_3 - \mathbb{Q}_2) & \dot{\mathbb{Q}}_3 = \frac{-\text{Å}g}{\Delta x}(\mathcal{H}_4 - \mathcal{H}_3) - \frac{\Finv}{\eth \text{Å}} \mathbb{Q}_3 \\ \dot{\mathcal{H}}_4 = \frac{-a^2}{g\text{Å}\Delta x}(\mathbb{Q}_4 - \mathbb{Q}_3) & \dot{\mathbb{Q}}_4 = \frac{-\text{Å}g}{\Delta x}(\mathcal{H}_5 - \mathcal{H}_4) - \frac{\Finv}{\eth \text{Å}} \mathbb{Q}_4 \\ \dot{\mathcal{H}}_5 = \frac{-a^2}{g\text{Å}\Delta x}(\mathbb{Q}_5 - \mathbb{Q}_4) & \end{array} \tag{3.66}$$

Therefore, the model $x = (Q_1 Q_2 Q_3 Q_4 \mathcal{H}_2 \mathcal{H}_3 \mathcal{H}_4)^T$, $u = (\mathcal{H}_1 \mathcal{H}_5)^T$, $y = (\mathbb{Q}_1 \mathbb{Q}_4)^T$ is defined as:

$$\begin{aligned} \dot{x}_1 &= \frac{-\text{Å}g}{\Delta x}(x_5 - u_1) - \frac{\Finv}{\eth \text{Å}} x_1 \\ \dot{x}_2 &= \frac{-\text{Å}g}{\Delta x}(x_6 - x_5) - \frac{\Finv}{\eth \text{Å}} x_2 \\ \dot{x}_3 &= \frac{-\text{Å}g}{\Delta x}(x_7 - x_6) - \frac{\Finv}{\eth \text{Å}} x_3 \\ \dot{x}_4 &= \frac{-\text{Å}g}{\Delta x}(u_2 - x_7) - \frac{\Finv}{\eth \text{Å}} x_4 \\ \dot{x}_5 &= \frac{-\text{Å}a^2}{g\text{Å}\Delta x}(x_2 - x_1) \\ \dot{x}_6 &= \frac{-a^2}{g\text{Å}\Delta x}(x_3 - x_2) \\ \dot{x}_7 &= \frac{-\text{Å}a^2}{g\text{Å}\Delta x}(x_4 - x_3) \end{aligned} \tag{3.67}$$

The variables A, B, C are defined as follows:

$$A = \begin{bmatrix} A_1 & A_2 \\ A_3 & A_4 \end{bmatrix}, \tag{3.68}$$

$$B = \begin{bmatrix} B_1 \\ B_2 \end{bmatrix}, \tag{3.69}$$

$$C = \begin{bmatrix} C_1 & C_2 \end{bmatrix} \tag{3.70}$$

$$A_1 = \begin{bmatrix} N & 0 & 0 & 0 \\ 0 & N & 0 & 0 \\ 0 & 0 & N & 0 \\ 0 & 0 & 0 & N \end{bmatrix}$$

$$A_2 = \begin{bmatrix} M & 0 & 0 & 0 \\ -M & M & 0 & 0 \\ 0 & -M & M & 0 \\ 0 & 0 & -M & M \end{bmatrix}$$

$$A_3 = \begin{bmatrix} S & 0 & 0 & 0 \\ -S & S & 0 & 0 \\ 0 & -S & S & 0 \\ 0 & 0 & -S & S \end{bmatrix}$$

$$A_4 = \begin{bmatrix} 0 & 0 & 0 & 0 \\ 0 & 0 & 0 & 0 \\ 0 & 0 & 0 & 0 \\ 0 & 0 & 0 & 0 \end{bmatrix} \tag{3.71}$$

$$B_1 = \begin{bmatrix} M & 0 \\ -M & 0 \\ 0 & 0 \\ 0 & M \end{bmatrix}$$

$$B_2 = \begin{bmatrix} 0 & 0 \\ 0 & 0 \\ 0 & 0 \\ 0 & 0 \end{bmatrix} \tag{3.72}$$

$$C_1 = \begin{bmatrix} 1 & 0 & 0 & 0 \\ 0 & 0 & 0 & 1 \end{bmatrix}$$

$$C_2 = \begin{bmatrix} 0 & 0 & 0 & 0 \\ 0 & 0 & 0 & 0 \end{bmatrix} \tag{3.73}$$

such that N, M, S are described as:

$$S = \frac{-a^2}{g\mathring{A}\Delta x}$$
$$N = -\frac{\Finv}{\eth\mathring{A}}$$
$$M = \frac{-Ag}{\Delta x}$$

3.4.2 Case 2

The initial pressure head at the first point of the pipe entrance and the flow rate at the end point of the pipe section are taken as boundary conditions. We have considered the following input conditions and assumptions,

$$\begin{cases} \mathbb{Q}(L,t) = \mathbb{Q}_{out}(t) \\ \mathcal{H}(0,t) = \mathcal{H}_{in}(t) \end{cases} \tag{3.74}$$

Also, we have considered the following output conditions and assumptions,

$$\begin{cases} \mathbb{Q}(0,t) = \mathbb{Q}_{in}(t) \\ \mathcal{H}(L,t) = \mathcal{H}_{out}(t) \end{cases} \tag{3.75}$$

In our designed model we have supposed that $x = (\mathbb{Q}_1\mathbb{Q}_2\mathbb{Q}_3\mathbb{Q}_4\mathcal{H}_2\mathcal{H}_3\mathcal{H}_4\mathcal{H}_5)^T$, $u[\mathbb{Q}_{out}\mathcal{H}_{in}]^T = (\mathbb{Q}_5\mathcal{H}_1)^T$ and $y = (\mathbb{Q}_1\mathcal{H}_5)^T$

$$\dot{\mathbb{Q}}_1 = \frac{-\mathring{A}g}{\Delta x}(\mathcal{H}_2 - \mathcal{H}_1) - \frac{\Finv}{\eth\mathring{A}}\mathbb{Q}_1$$
$$\dot{\mathbb{Q}}_2 = \frac{-\mathring{A}g}{2\Delta x}(\mathcal{H}_3 - \mathcal{H}_1) - \frac{\Finv}{\eth\mathring{A}}\mathbb{Q}_2$$
$$\dot{\mathbb{Q}}_3 = \frac{-\mathring{A}g}{2\Delta x}(\mathcal{H}_4 - \mathcal{H}_2) - \frac{\Finv}{\eth\mathring{A}}\mathbb{Q}_3$$
$$\dot{\mathbb{Q}}_4 = \frac{-\mathring{A}g}{2\Delta x}(\mathcal{H}_5 - \mathcal{H}_3) - \frac{\Finv}{\eth\mathring{A}}\mathbb{Q}_4$$
$$\dot{\mathcal{H}}_2 = \frac{-a^2}{2g\Delta x}(\mathbb{Q}_3 - \mathbb{Q}_1)$$
$$\dot{\mathcal{H}}_3 = \frac{-a^2}{2g\Delta x}(\mathbb{Q}_4 - \mathbb{Q}_2)$$

$$\dot{\mathcal{H}}_4 = \frac{-a^2}{2g\Delta x}(\mathbb{Q}_5 - \mathbb{Q}_3)$$
$$\dot{\mathcal{H}}_5 = \frac{-a^2}{g\Delta x}(\mathbb{Q}_5 - \mathbb{Q}_4) \tag{3.76}$$

The variables A, B, C in (3.61) and (3.65) are defined as follows:

$$A = \begin{bmatrix} N & 0 & 0 & 0 & 2M & 0 & 0 & 0 \\ 0 & N & 0 & 0 & 0 & M & 0 & 0 \\ 0 & 0 & N & 0 & -M & 0 & M & 0 \\ 0 & 0 & 0 & N & 0 & -M & 0 & M \\ -S & 0 & S & 0 & 0 & 0 & 0 & 0 \\ 0 & -S & 0 & S & 0 & 0 & 0 & 0 \\ 0 & 0 & -S & 0 & 0 & 0 & 0 & 0 \\ 0 & 0 & 0 & -S & 0 & 0 & 0 & 0 \end{bmatrix}$$

$$B = \begin{bmatrix} 0 & 2M \\ 0 & M \\ 0 & 0 \\ 0 & 0 \\ 0 & 0 \\ 0 & 0 \\ S & 0 \\ 2S & 0 \end{bmatrix} \quad \textit{and}$$

$$C = \begin{bmatrix} 1 & 0 & 0 & 0 & 0 & 0 & 0 & 0 \\ 0 & 0 & 0 & 0 & 0 & 0 & 0 & 1 \end{bmatrix} \tag{3.77}$$

such that N, M, S are described as:

$$S = \frac{-a^2}{2g\text{Å}\Delta x}$$
$$N = -\frac{Ⅎ}{\text{ðÅ}}$$
$$M = \frac{-Ag}{2\Delta x} \tag{3.78}$$

3.4.3 Case 3

The flow rate and pressure head at the first point of the pipe entrance are taken as boundary conditions. We have considered the following input conditions and assumptions,

$$\begin{cases} \mathbb{Q}(0,t) = \mathbb{Q}_{in}(t) \\ H(0,t) = \mathcal{H}_{in}(t) \end{cases} \tag{3.79}$$

Also, we have considered the following output conditions and assumptions,

$$\begin{cases} \mathbb{Q}(\mathrm{L},t) = \mathbb{Q}_{out}(t) \\ \mathcal{H}(L,t) = \mathcal{H}_{out}(t) \end{cases} \tag{3.80}$$

In our designed model we have supposed that $x = (\mathbb{Q}_2\mathbb{Q}_3\mathbb{Q}_4\mathbb{Q}_5\mathcal{H}_2\mathcal{H}_3\mathcal{H}_4\mathcal{H}_5)^T$, $u[\mathbb{Q}_{in}\mathcal{H}_{in}]^T = (\mathbb{Q}_1\mathcal{H}_1)^T$ and $y = (\mathbb{Q}_5\mathcal{H}_5)^T$.

$$\begin{aligned}
\dot{\mathbb{Q}}_2 &= \frac{-g}{2\Delta x}(\mathcal{H}_3 - \mathcal{H}_1) - \frac{\text{Ⅎ}}{\eth\mathring{A}}\mathbb{Q}_2 \\
\dot{\mathbb{Q}}_3 &= \frac{-g}{2\Delta x}(\mathcal{H}_4 - \mathcal{H}_2) - \frac{\text{Ⅎ}}{\eth\mathring{A}}\mathbb{Q}_3 \\
\dot{\mathbb{Q}}_4 &= \frac{-g}{2\Delta x}(\mathcal{H}_5 - \mathcal{H}_3) - \frac{\text{Ⅎ}}{\eth\mathring{A}}\mathbb{Q}_4 \\
\dot{\mathbb{Q}}_5 &= \frac{-g}{\Delta x}(\mathcal{H}_5 - \mathcal{H}_4) - \frac{\text{Ⅎ}}{\eth\mathring{A}}\mathbb{Q}_5 \\
\dot{\mathcal{H}}_2 &= \frac{-a^2}{2g\mathring{A}\Delta x}(\mathbb{Q}_3 - \mathbb{Q}_1) \\
\dot{\mathcal{H}}_3 &= \frac{-a^2}{2g\mathring{A}\Delta x}(\mathbb{Q}_4 - \mathbb{Q}_2) \\
\dot{\mathcal{H}}_4 &= \frac{-a^2}{2g\mathring{A}\Delta x}(\mathbb{Q}_5 - \mathbb{Q}_3) \\
\dot{\mathcal{H}}_5 &= \frac{-a^2}{g\Delta x}(\mathbb{Q}_5 - \mathbb{Q}_4)
\end{aligned} \tag{3.81}$$

The variables A, B, C in (3.61) and (3.65) are defined as follows:

$$
A = \begin{bmatrix}
N & 0 & 0 & 0 & 0 & -M & 0 & 0 \\
0 & N & 0 & 0 & 0 & 0 & -M & 0 \\
0 & 0 & N & 0 & M & 0 & 0 & -M \\
0 & 0 & 0 & N & 0 & M & 0 & -M \\
0 & -S & 0 & 0 & 0 & 0 & 0 & 0 \\
S & 0 & -S & 0 & 0 & 0 & 0 & 0 \\
0 & S & 0 & -S & 0 & 0 & 0 & 0 \\
0 & 0 & S & 0 & 0 & 0 & 0 & 0
\end{bmatrix}
$$

$$
B = \begin{bmatrix}
0 & M \\
0 & 0 \\
0 & 0 \\
0 & 0 \\
S & 0 \\
0 & 0 \\
0 & 0 \\
0 & 0
\end{bmatrix} \quad and
$$

$$
C = \begin{bmatrix}
0 & 0 & 0 & 1 & 0 & 0 & 0 & 0 \\
0 & 0 & 0 & 0 & 1 & 0 & 0 & 0
\end{bmatrix} \tag{3.82}
$$

such that N, M, S are described as:

$$
\begin{aligned}
S &= \frac{-a^2}{2g\mathring{A}\Delta x} \\
N &= -\frac{\Finv}{\eth\mathring{A}} \\
M &= \frac{-g}{2\Delta x}
\end{aligned} \tag{3.83}
$$

3.4.4 Case 4

The initial flow rate at the first point of the pipe entrance and the pressure head at the end point of the pipe section are taken as boundary conditions. We have considered the following input conditions and assumptions,

$$
\begin{cases}
\mathbb{Q}(0, t) = \mathbb{Q}_{in}(t) \\
\mathcal{H}(L, t) = \mathcal{H}_{out}(t)
\end{cases} \tag{3.84}
$$

Also, we have considered the following output conditions and assumptions,

$$\begin{cases} \mathcal{H}(0,t) = \mathcal{H}_{in}(t) \\ \mathbb{Q}(L,t) = \mathbb{Q}_{out}(t) \end{cases} \tag{3.85}$$

In our designed model we have supposed that $x = (\mathbb{Q}_2\mathbb{Q}_3\mathbb{Q}_4\mathbb{Q}_5\mathcal{H}_1\mathcal{H}_2\mathcal{H}_3\mathcal{H}_4)^T$, $u[\mathbb{Q}_{in}\mathcal{H}_{out}]^T = (\mathbb{Q}_1\mathcal{H}_5)^T$ and $y = (\mathbb{Q}_5\mathcal{H}_1)^T$.

$$\begin{aligned}
\dot{\mathbb{Q}}_2 &= \frac{-\mathring{A}g}{2\Delta x}(\mathcal{H}_3 - \mathcal{H}_1) - \frac{\Finv}{\eth\mathring{A}}\mathbb{Q}_2 \\
\dot{\mathbb{Q}}_3 &= \frac{-\mathring{A}g}{2\Delta x}(\mathcal{H}_4 - \mathcal{H}_2) - \frac{\Finv}{\eth\mathring{A}}\mathbb{Q}_3 \\
\dot{\mathbb{Q}}_4 &= \frac{-\mathring{A}g}{2\Delta x}(\mathcal{H}_5 - \mathcal{H}_3) - \frac{\Finv}{\eth\mathring{A}}\mathbb{Q}_4 \\
\dot{\mathbb{Q}}_5 &= \frac{-\mathring{A}g}{\Delta x}(\mathcal{H}_5 - \mathcal{H}_4) - \frac{\Finv}{\eth\mathring{A}}\mathbb{Q}_5 \\
\dot{\mathcal{H}}_1 &= \frac{-a^2}{2g\mathring{A}\Delta x}(\mathbb{Q}_2 - \mathbb{Q}_1) \\
\dot{\mathcal{H}}_2 &= \frac{-a^2}{2g\mathring{A}\Delta x}(\mathbb{Q}_3 - \mathbb{Q}_1) \\
\dot{\mathcal{H}}_3 &= \frac{-a^2}{2g\mathring{A}\Delta x}(\mathbb{Q}_4 - \mathbb{Q}_2) \\
\dot{\mathcal{H}}_4 &= \frac{-a^2}{2g\mathring{A}\Delta x}(\mathbb{Q}_5 - \mathbb{Q}_3)
\end{aligned} \tag{3.86}$$

The variables A, B, C in (3.61) and (3.65) are defined as follows:

$$A = \begin{bmatrix} N & 0 & 0 & 0 & -M & 0 & M & 0 \\ 0 & N & 0 & 0 & 0 & -M & 0 & M \\ 0 & 0 & N & 0 & 0 & 0 & -M & 0 \\ 0 & 0 & 0 & N & 0 & 0 & 0 & -M \\ 2S & 0 & 0 & 0 & 0 & 0 & 0 & 0 \\ 0 & S & 0 & 0 & 0 & 0 & 0 & 0 \\ -S & 0 & S & 0 & 0 & 0 & 0 & 0 \\ 0 & -S & 0 & S & 0 & 0 & 0 & 0 \end{bmatrix}$$

$$B = \begin{bmatrix} 0 & 0 \\ 0 & 0 \\ 0 & M \\ 0 & -2M \\ -2S & 0 \\ S & 0 \\ 0 & 0 \\ 0 & 0 \end{bmatrix} \quad and$$

$$C = \begin{bmatrix} 0\,0\,0\,1\,0\,0\,0\,0 \\ 0\,0\,0\,0\,1\,0\,0\,0 \end{bmatrix} \tag{3.87}$$

such that N, M, S are described as:

$$\begin{aligned} S &= \frac{-a^2}{2g\mathring{A}\Delta x} \\ N &= -\frac{\Finv}{\eth\mathring{A}} \\ M &= \frac{-g}{2\Delta x} \end{aligned} \tag{3.88}$$

3.4.5 *Case 5*

The flow rate at the first point of the pipe entrance and the end point of the pipe section are taken as initial and boundary conditions. We have considered the following input conditions and assumptions,

$$\begin{cases} \mathbb{Q}(0,t) = \mathbb{Q}_{in}(t) \\ \mathbb{Q}(L,t) = \mathbb{Q}_{out}(t) \end{cases} \tag{3.89}$$

Also, we have considered the following output conditions and assumptions,

$$\begin{cases} \mathcal{H}(0,t) = \mathcal{H}_{in}(t) \\ \mathcal{H}(L,t) = \mathcal{H}_{out}(t) \end{cases} \tag{3.90}$$

In our designed model we have supposed that $x = (\mathbb{Q}_2\mathbb{Q}_3\mathbb{Q}_4\mathcal{H}_1\mathcal{H}_2\mathcal{H}_3\mathcal{H}_4\mathcal{H}_5)^T$, $u[\mathbb{Q}_{in}\mathbb{Q}_{out}]^T = (\mathbb{Q}_1\mathbb{Q}_5)^T$ and $y = (\mathcal{H}_1\mathcal{H}_5)^T$.

$$\dot{\mathbb{Q}}_2 = \frac{-\mathring{A}g}{2\Delta x}(\mathcal{H}_3 - \mathcal{H}_1) - \frac{\Finv}{\eth\mathring{A}}\mathbb{Q}_2$$

$$
\begin{aligned}
\dot{\mathbb{Q}}_3 &= \frac{-\mathring{A}g}{2\Delta x}(\mathcal{H}_4 - \mathcal{H}_2) - \frac{\Finv}{\eth\mathring{A}}\mathbb{Q}_3 \\
\dot{\mathbb{Q}}_4 &= \frac{-\mathring{A}g}{2\Delta x}(\mathcal{H}_5 - \mathcal{H}_3) - \frac{\Finv}{\eth\mathring{A}}\mathbb{Q}_4 \\
\dot{\mathcal{H}}_1 &= \frac{-a^2}{g\mathring{A}\Delta x}(\mathbb{Q}_2 - \mathbb{Q}_1) \\
\dot{\mathcal{H}}_2 &= \frac{-a^2}{2g\mathring{A}\Delta x}(\mathbb{Q}_3 - \mathbb{Q}_1) \\
\dot{\mathcal{H}}_3 &= \frac{-a^2}{2g\mathring{A}\Delta x}(\mathbb{Q}_4 - \mathbb{Q}_2) \\
\dot{\mathcal{H}}_4 &= \frac{-a^2}{2g\mathring{A}\Delta x}(\mathbb{Q}_5 - \mathbb{Q}_3) \\
\dot{\mathcal{H}}_5 &= \frac{-a^2}{g\mathring{A}\Delta x}(\mathbb{Q}_5 - \mathbb{Q}_4)
\end{aligned}
\tag{3.91}
$$

The variables A, B, C in (3.61) and (3.65) are defined as follows:

$$
A = \begin{bmatrix}
N & 0 & 0 & -M & 0 & M & 0 & 0 \\
0 & N & 0 & 0 & -M & 0 & M & 0 \\
0 & 0 & N & 0 & 0 & -M & 0 & M \\
2S & 0 & 0 & 0 & 0 & 0 & 0 & 0 \\
0 & S & 0 & 0 & 0 & 0 & 0 & 0 \\
-S & 0 & S & 0 & 0 & 0 & 0 & 0 \\
0 & -S & 0 & 0 & 0 & 0 & 0 & 0 \\
0 & 0 & -S & 0 & 0 & 0 & 0 & 0
\end{bmatrix}
$$

$$
B = \begin{bmatrix}
0 & 0 \\
0 & 0 \\
0 & 0 \\
-2S & 0 \\
-S & 0 \\
0 & 0 \\
0 & S \\
0 & 2S
\end{bmatrix} \quad \textit{and}
$$

$$
c = \begin{bmatrix} 0\,0\,0\,1\,0\,0\,0\,0 \\ 0\,0\,0\,0\,0\,0\,0\,1 \end{bmatrix} \tag{3.92}
$$

such that N, M, S are described as:

$$
S = \frac{-a^2}{2g\mathring{A}\Delta x}
$$

$$
\begin{aligned}
N &= -\frac{Ⅎ}{\eth \mathring{A}} \\
M &= \frac{-\mathring{A}g}{2\Delta x}
\end{aligned}
\tag{3.93}
$$

3.4.6 *Case 6*

The flow rate and pressure head at the end point of the pipe section are taken as initial and boundary conditions. We have considered the following input conditions and assumptions,

$$
\begin{cases}
\mathbb{Q}(\mathrm{L}, t) = \mathbb{Q}_{out}(t) \\
\mathcal{H}(L, t) = \mathcal{H}_{out}(t)
\end{cases}
\tag{3.94}
$$

Also, we have considered the following output conditions and assumptions,

$$
\begin{cases}
\mathbb{Q}(0, t) = \mathbb{Q}_{in}(t) \\
\mathcal{H}(0, t) = \mathcal{H}_{in}(t)
\end{cases}
\tag{3.95}
$$

In our designed model we have supposed that $x = (\mathbb{Q}_1\mathbb{Q}_2\mathbb{Q}_3\mathbb{Q}_4\mathcal{H}_1\mathcal{H}_2\mathcal{H}_3\mathcal{H}_4)^T$, $u[\mathbb{Q}_{out}\mathcal{H}_{out}]^T = (\mathbb{Q}_5\mathcal{H}_5)^T$ and $y = (\mathbb{Q}_1\mathcal{H}_1)^T$.

$$
\begin{aligned}
\dot{\mathbb{Q}}_1 &= \frac{-\mathring{A}g}{\Delta x}(\mathcal{H}_2 - \mathcal{H}_1) - \frac{Ⅎ}{\eth \mathring{A}}\mathbb{Q}_1 \\
\dot{\mathbb{Q}}_2 &= \frac{-\mathring{A}g}{2\Delta x}(\mathcal{H}_3 - \mathcal{H}_1) - \frac{Ⅎ}{\eth \mathring{A}}\mathbb{Q}_2 \\
\dot{\mathbb{Q}}_3 &= \frac{-\mathring{A}g}{2\Delta x}(\mathcal{H}_4 - \mathcal{H}_2) - \frac{Ⅎ}{\eth \mathring{A}}\mathbb{Q}_3 \\
\dot{\mathbb{Q}}_4 &= \frac{-\mathring{A}g}{2\Delta x}(\mathcal{H}_5 - \mathcal{H}_3) - \frac{Ⅎ}{\eth \mathring{A}}\mathbb{Q}_4 \\
\dot{\mathcal{H}}_1 &= \frac{-a^2}{g\mathring{A}\Delta x}(\mathbb{Q}_2 - \mathbb{Q}_1) \\
\dot{\mathcal{H}}_2 &= \frac{-a^2}{2g\mathring{A}\Delta x}(\mathbb{Q}_3 - \mathbb{Q}_1) \\
\dot{\mathcal{H}}_3 &= \frac{-a^2}{2g\mathring{A}\Delta x}(\mathbb{Q}_4 - \mathbb{Q}_2) \\
\dot{\mathcal{H}}_4 &= \frac{-a^2}{2g\mathring{A}\Delta x}(\mathbb{Q}_5 - \mathbb{Q}_3)
\end{aligned}
\tag{3.96}
$$

The variables A, B, C in (3.61) and (3.65) are defined as follows:

$$A = \begin{bmatrix} N & 0 & 0 & 0 & -2M & 2M & 0 & 0 \\ 0 & N & 0 & 0 & -M & 0 & M & 0 \\ 0 & 0 & N & 0 & 0 & -M & 0 & M \\ 0 & 0 & 0 & N & 0 & 0 & -M & 0 \\ -2S & 2S & 0 & 0 & 0 & 0 & 0 & 0 \\ -S & 0 & S & 0 & 0 & 0 & 0 & 0 \\ 0 & -S & 0 & S & 0 & 0 & 0 & 0 \\ 0 & 0 & -S & 0 & 0 & 0 & 0 & 0 \end{bmatrix}$$

$$B = \begin{bmatrix} 0 & 0 \\ 0 & 0 \\ 0 & 0 \\ 0 & M \\ 0 & 0 \\ 0 & 0 \\ 0 & 0 \\ S & 0 \end{bmatrix} \quad \textit{and}$$

$$c = \begin{bmatrix} 1 & 0 & 0 & 0 & 0 & 0 & 0 & 0 \\ 0 & 0 & 0 & 0 & 1 & 0 & 0 & 0 \end{bmatrix} \tag{3.97}$$

such that N, M, S are described as:

$$\begin{aligned} S &= \frac{-a^2}{2g\text{Å}\Delta x} \\ N &= -\frac{⅃}{\eth\text{Å}} \\ M &= \frac{-\text{Å}g}{2\Delta x} \end{aligned} \tag{3.98}$$

3.5 Observability and Controllability Analysis of Linear System

The aim of this section is to test the controllability as well as the observability of the linear system (3.56) using the following Lemmas.

Lemma 1 *Let us consider the linear system* (3.56) *having* N_u *control inputs (pressure head), and* N_y *outputs (flow rates). The controllability of matrix* $\mathbb{C}_i$ *can be described as follows* [66]:

$$\mathbb{C}_i(A, B)\left[BABA^2BA^3B\ldots A^{i-1}B\right] \tag{3.99}$$

Applying the Cayley-Hamilton theorem [67], *the rank of* $\mathbb{C}_i$ *can be described using the first* $N \times N_u$*columns, in which* N*is taken to be the state dimension which is the same as with the dimension of Matrix A.*

The matrix pair (A, B)*is known controllable when its controllability matrix contains a full row rank.*

Lemma 2 *For linear system* (3.56) *the observability of matrix* O_i *can be described as follows* [68]:

$$\mathrm{O}_i(C, A) := \begin{bmatrix} C \\ CA \\ CA^2 \\ \vdots \\ CA^{i-1} \end{bmatrix} \tag{3.100}$$

Applying the Cayley-Hamilton theorem, the rank of O_i *can be described using the first* $N \times N_u$*rows, in which N is taken to be the state dimension which is the same as with the dimension of Matrix A the matrix pair* (A, C)*is known observable when its observability matrix has a full column rank.*

References

1. Billmann, L., Isermann, R.: Leak detection methods for pipelines. Automatica **23**(3), 381–385 (1987)
2. Kim, S.H.: Extensive development of leak detection algorithm by impulse response method. J. Hydraul. Eng. **131**(3), 201–208 (2005)
3. Loparo, K.A., Buchner, M., Vasudeva, K.S.: Leak detection in an experimental heat exchanger process: a multiple model approach. IEEE Trans. Autom. Control **36**(2), 167–177 (1991)
4. Xie, Y.: Aiken A Context-and path-sensitive memory leak detection. In: Proceedings of the 10th European Software Engineering Conference Held Jointly with 13th ACM SIGSOFT International Symposium on Foundations of Software Engineering, pp 115–125 (2005)
5. Cherem, S., Princehouse, L., Rugina, R.: Practical memory leak detection using guarded value-flow analysis. In: Proceedings of the 28th ACM SIGPLAN Conference on Programming Language Design and Implementation, pp. 480–491 (2007)
6. Greene, D.A., Greene, R.A., Gaubatz, D.C.: Integrated acoustic leak detection processing system. Google Patents (1996)
7. Begovich, O., Pizano-Moreno, A.: Application of a leak detection algorithm in a water pipeline prototype: Difficulties and solutions. In: 2008 5th International Conference on Electrical Engineering, Computing Science and Automatic Control, pp. 26–30. IEEE (2008)
8. Sun, X., Chen, T., Marquez, H.J.: Efficient model-based leak detection in boiler steam-water systems. Comput. Chem. Eng. **26**(11), 1643–1647 (2002)
9. Hou, M., Müller, P.C.: Fault detection and isolation observers. Int. J. Control **60**(5), 827–846 (1994)

10. Salvesen, J.: Leak Detection by Estimation in an Oil Pipeline. Norwegian University of Science and Technology (NTNU), Trondheim, Norway, MScthesis (2005)
11. Verde, C.: Minimal order nonlinear observer for leak detection. J. Dyn. Sys. Meas. Control **126**(3), 467–472 (2004)
12. Verde, C.: Multi-leak detection and isolation in fluid pipelines. Control Eng. Pract. **9**(6), 673–682 (2001)
13. Yang, J., Qingxin, Y., Guanghai, L., Jingyan, Z.: Acoustic emission source identification technique for buried gas pipeline leak. In: 2006 9th International Conference on Control, Automation, Robotics and Vision, pp. 1–5. IEEE (2006)
14. Lee, N.Y., Hwang, I.S., Yoo, H.I.: New leak detection technique using ceramic humidity sensor for water reactors. Nucl. Eng. Des. **205**(1–2), 23–33 (2001)
15. (!!! INVALID CITATION !!! [14–16]).
16. Scalapino, D., Imry, Y., Pincus, P.: Generalized Ginzburg-Landau theory of pseudo-one-dimensional systems. Phys. Rev. B **11**(5), 2042 (1975)
17. Abrahams, E., Tsuneto, T.: Time variation of the Ginzburg-Landau order parameter. Phys. Rev. **152**(1), 416 (1966)
18. Yu, W., Jafari, R.: Modeling and Control of Uncertain Nonlinear Systems with Fuzzy Equations and Z-Number. Wiley (2019)
19. Razvarz, S., Jafari, R.: ICA and ANN modeling for photocatalytic removal of pollution in wastewater. Math. Comput. Appl. **22**(3), 38 (2017a)
20. Jafari, R., Razvarz, S., Gegov, A.: Neural network approach to solving fuzzy nonlinear equations using Z-numbers. IEEE Trans. Fuzzy Syst. (2019)
21. Razvarz, S., Jafari, R.: Intelligent techniques for photocatalytic removal of pollution in wastewater. J. Electr. Eng. **5**(1), 321–328 (2017b)
22. Jafari, R., Razvarz, S., Gegov, A., Paul, S., Keshtkar, S.: Fuzzy Sumudu transform approach to solving fuzzy differential equations with Z-numbers. In: Advanced Fuzzy Logic Approaches in Engineering Science, pp. 18–48. IGI Global (2019)
23. Jafari, R., Razvarz, S., Gegov, A.: A novel technique to solve fully fuzzy nonlinear matrix equations. In: International Conference on Theory and Applications of Fuzzy Systems and Soft Computing, 2018, pp. 886–892. Springer (2018)
24. Jafari, R., Razvarz, S., Gegov, A.: Fuzzy differential equations for modeling and control of fuzzy systems. In: International Conference on Theory and Applications of Fuzzy Systems and Soft Computing, 2018, pp. 732–740. Springer (2018)
25. Jafari, R., Yu, W., Razvarz, S., Gegov, A.: Numerical methods for solving fuzzy equations: A survey. Fuzzy Sets Syst. (2019)
26. Jafari, R., Razvarz, S., Gegov, A.: A new computational method for solving fully fuzzy nonlinear systems. In: International Conference on Computational Collective Intelligence, 2018, pp. 503–512. Springer (2018)
27. Jafari, R., Razvarz, S., Gegov, A., Paul, S.: Modeling and control of uncertain nonlinear systems. In: 2018 International Conference on Intelligent Systems (IS), pp. 168–173. IEEE (2018)
28. Jafari, R., Razvarz, S., Gegov, A.: A novel technique for solving fully fuzzy nonlinear systems based on neural networks. Vietnam J. Comput. Sci. **7**(1), 93–107 (2020)
29. Razvarz, S., Hernández-Rodríguez, F., Jafari, R., Gegov, A.: Foundation of Z-Numbers and Engineering Applications. In: Latin American Symposium on Industrial and Robotic Systems, pp. 15–24. Springer (2019)
30. Jafari, R., Contreras, M.A., Yu, W., Gegov, A.: Applications of fuzzy logic, artificial neural network and neuro-fuzzy in industrial engineering. In: Latin American Symposium on Industrial and Robotic Systems, pp. 9–14. Springer (2019)
31. Jafari, R., Razvarz, S., Gegov, A., Yu, W.: Fuzzy control of uncertain nonlinear systems with numerical techniques: a survey. In: UK Workshop on Computational Intelligence, pp. 3–14. Springer (2019)
32. Jafari, R., Razvarz, S., Yu, W., Gegov, A., Goodwin, M., Adda, M.: Genetic algorithm modeling for photocatalytic elimination of impurity in wastewater. In: Proceedings of SAI Intelligent Systems Conference, pp. 228–236. Springer (2019)

33. Tatchum, M., Gegov, A., Jafari, R., Razvarz, S.: Parallel distributed compensation for voltage controlled active magnetic bearing system using integral fuzzy model. In: 2018 International Conference on Intelligent Systems (IS), pp. 190–198. IEEE (2018)
34. Razvarz, S., Jafari, R., Gegov, A.: Solving partial differential equations with Bernstein neural networks. In: UK Workshop on Computational Intelligence, pp. 57–70. Springer (2018)
35. Jafarian, A., Jafari, R.: New iterative approach for solving fully fuzzy polynomials. Int. J. Fuzzy Math. Syst. **3**(2):75–83
36. Jafarian, A., Jafari, R.: New method for solving fuzzy polynomials. Adv. Fuzzy Math. **8**(1), 25–33 (2013)
37. Jafarian, A., Jafari, R.: An iterative method for solving fuzzy polynomials by fuzzy neural networks (2012)
38. Jafarian, A., Jafari, R.: Simulation and evaluation of fuzzy polynomials by feed-back neural networks (2012)
39. Jafari, R., Yu, W.: Fuzzy control for uncertainty nonlinear systems with dual fuzzy equations. J. Intell. Fuzzy Syst. **29**(3), 1229–1240 (2015)
40. Jafari, R., Yu, W.: Fuzzy modeling for uncertainty nonlinear systems with fuzzy equations. Math. Prob. Eng. (2017)
41. Razvarz, S., Vargas-Jarillo, C., Jafari, R.: Pipeline monitoring architecture based on observability and controllability analysis. In: 2019 IEEE International Conference on Mechatronics (ICM), 18–20 March 2019, pp. 420–423
42. Jafari, R., Razvarz, S., Vargas-Jarillo, C., Yu, W.: Control of flow rate in pipeline using PID controller. In: 2019 IEEE 16th International Conference on Networking, Sensing and Control (ICNSC), 9–11 May 2019, pp. 293–298
43. Jafari, R.R., S., Vargas-Jarillo, C., Gegov, A, : Blockage detection in pipeline based on the extended Kalman filter observer. Electronics **9**(1), 91–107 (2020)
44. Razvarz, S., Jafari, R., Vargas-Jarillo, C.: Modelling and analysis of flow rate and pressure head in pipelines. In: 2019 16th International Conference on Electrical Engineering, Computing Science and Automatic Control (CCE), pp. 1–6. IEEE (2019)
45. Jafari, R., Razvarz, S., Vargas-Jarillo, C., Gegov, A.E.: The effect of baffles on heat transfer. In: ICINCO, vol. 2, pp 607–612 (2019)
46. Razvarz, S., Jafari, R., Vargas-Jarillo, C., Gegov, A., Forooshani, M.: Leakage detection in pipeline based on second order extended Kalman filter observer. IFAC-PapersOnLine **52**(29), 116–121 (2019)
47. Razvarz, S., Vargas-Jarillo, C., Jafari, R., Gegov, A.: Flow control of fluid in pipelines using PID controller. IEEE Access **7**, 25673–25680 (2019)
48. Razvarz, S., Chavez, L.F.G., Vargas-Jarillo, C.: Nanotechnology Applications in Industry and Heat Transfer. In: Latin American Symposium on Industrial and Robotic Systems, pp. 1–8. Springer (2019)
49. Batchelor, C.K., Batchelor, G.: An Introduction to Fluid Dynamics. Cambridge University Press (2000)
50. Landau, L., Lifshitz, E.: Fluid Mechanics, V. 6 of Course of Theoretical Physics, 2nd English edition. Revised. Pergamon Press, Oxford-New York-Beijing-Frankfurt-San Paulo-Sydney-Tokyo-Toronto (1987)
51. Bedford, A., Fowler, W.: Engineering Mechanics: Statics and Dynamics. Prentice-Hall (2008)
52. Shames, I.H., Rao, G.K.M.: Engineering Mechanics: Statics and Dynamics. Prentice-Hall Englewood Cliffs (1967)
53. Falkovich, G.: Fluid mechanics: A Short Course for Physicists. Cambridge University Press (2011)
54. Whitaker, R.D.: An historical note on the conservation of mass. J. Chem. Educ. **52**(10), 658 (1975)
55. Bolton, P., Thatcher, R.: On mass conservation in least-squares methods. J. Comput. Phys. **203**(1), 287–304 (2005)
56. Kattelans, T., Heinrichs, W.: Conservation of mass and momentum of the least-squares spectral collocation scheme for the Stokes problem. J. Comput. Phys. **228**(13), 4649–4664 (2009)

57. DeLong, L.L.: Mass conservation: 1-D open channel flow equations. J. Hydraul. Eng. **115**(2), 263–269 (1989)
58. Whitham, G.: Mass, momentum and energy flux in water waves. J. Fluid Mech. **12**(1), 135–147 (1962)
59. Lee, J.R., Chung, S.Y.: Leibniz's rule for anti-differentiation. Int. J. Math. Educ. Sci. Technol. **22**(4), 645–650 (1991)
60. Cha, S., Kang, K., You, J.B., Im, S.G., Kim, Y., Kim, J.M.: Hoop stress-assisted three-dimensional particle focusing under viscoelastic flow. Rheol. Acta **53**(12), 927–933 (2014)
61. Chaudhry, M.H.: Applied hydraulic transients (1979)
62. Wylie, E.B., Streeter, V.L., Suo, L.: Fluid Transients in Systems, vol. 1. Prentice Hall Englewood Cliffs, NJ (1993)
63. Yan, P., Sekar, A.: Study of linear models in steady state load flow analysis of power systems. In: 2002 IEEE Power Engineering Society Winter Meeting. Conference Proceedings (Cat. No. 02CH37309), pp. 666–671. IEEE (2002)
64. Chopra, A.K., Carter, R.D.: Proof of the two-phase steady-state theory for flow through porous media. SPE Formation Eval. **1**(06), 603–608 (1986)
65. Manheimer, W.M., Colombant, D., Gardner, J.: Steady-state planar ablative flow. Phys. Fluids **25**(9), 1644–1652 (1982)
66. Khalil, H.K., Grizzle, J.W.: Nonlinear Systems, vol. 3. Prentice hall Upper Saddle River, NJ (2002)
67. Mertzios, B., Christodoulou, M.: On the generalized Cayley-Hamilton theorem. IEEE Trans. Autom. Control **31**(2), 156–157 (1986)
68. Chen, C.-T.: Linear System Theory and Design. Oxford University Press, Inc. (1998)

Chapter 4
Theory and Applications of Fuzzy Logic Controller for Flowing Fluids

4.1 Mathematical Preliminaries

Definition 1 (*Fuzzy Variable*) Assume that d is:

1. normal, there exists $\tau_0 \in \Re$ such that $d(\tau_0) = 1$,
2. convex,
3. upper semi-continuous on $\Re, d(\tau) \leq d(\tau_0) + \epsilon, \forall \tau \in N(\tau_0), \forall \tau_0 \in \Re, \forall \epsilon > 0$, $N(\tau_0)$ is a neighborhood,
4. is compact, hence d is a fuzzy variable, $d \in E : \Re \to [0, 1]$.
   The fuzzy variable d can be stated as below

$$\underline{d}, \overline{d} \tag{4.1}$$

where $\underline{d}$ is taken to be the lower-bound variable and $\overline{d}$ is taken to be the upper-bound variable [1–6].

Definition 2 (*number*)$\widehat{\mathbf{Z}}$ The $\widehat{Z}$-number can be defined as$\widehat{Z} = [d(\tau), \widetilde{p}]$, where is taken to be the restriction on τ, and $\widetilde{p}$ is taken to be the measure of the reliability of d [7–12]. $\widetilde{p}$ is the reliability, strength of belief, probability or possibility. The $\widehat{Z}$-number may be defined in the form of $\widehat{Z}^+$ -number, such that $d(\tau)$ is a fuzzy number and $\widetilde{p}$ is the probability distribution of τ. The $\widehat{Z}^-$ -number may be defined in the form of $\widehat{Z}^-$ -number, such that $d(\tau)$, and $\widetilde{p}$, are fuzzy numbers [13–15].

The $\widehat{Z}^+$ -number holds more information in comparison with the $\widehat{Z}^-$ -number. In this chapter, the $\widehat{Z}^+$ -number is employed, such that $\widehat{Z} = [d, \widetilde{p}]$, d is taken to be a fuzzy number and $\widetilde{p}$ is taken to be a probability distribution [16, 17].

The triangular membership function of a fuzzy number d can be defined as [18–20]

S. Razvarz et al., *Flow Modelling and Control in Pipeline Systems*, Studies in Systems, Decision and Control 321, https://doi.org/10.1007/978-3-030-59246-2_4

$$\mu_d = Q(h,k,l) = \begin{cases} \frac{\tau-h}{k-h} h \le \tau \le k \\ \frac{l-\tau}{l-k} k \le \tau \le l \end{cases} \quad otherwise \quad \mu_d = 0 \tag{4.2}$$

and the trapezoidal membership function of a fuzzy number d can be defined as

$$\mu_d = Q(h,k,l,s) = \begin{cases} \frac{\tau-h}{k-h} h \le \tau \le k \\ \frac{s-\tau}{s-l} l \le \tau \le s \quad otherwise \quad \mu_d = 0 \\ 1 k \le \tau \le l \end{cases} \tag{4.3}$$

In case d is a fuzzy event in $\Re$, the probability measure can be described as follows

$$P(d) = \int_{\Re} \mu_d(\tau)\tilde{p}(\tau)d\tau \tag{4.4}$$

such that $\tilde{p}$ is taken to be the probability density of τ. The probability measure for discrete $\widehat{Z}$-numbers can be described as follows

$$P(d) = \sum_{j=1}^{n} \mu_d(\tau_j)\ \tilde{p}\ (\tau_j) \tag{4.5}$$

Definition 3 (*r-level*) The r -level of the $\widehat{Z}$-number $\widehat{Z} = (d, \tilde{p})$ can be stated as below

$$[\widehat{Z}]^r = ([d]^r, [\tilde{p}]^r) \tag{4.6}$$

such that $0 < r \le 1$. $[\tilde{p}]^r$ can be calculated using the Nguyen's theorem

$$[\tilde{p}]^r = \tilde{p}\ ([d]^r) = \tilde{p}\ \left(\left[\underline{d}^r, \overline{d}^r\right]\right) = [\underline{P}^r, \overline{P}^r] \tag{4.7}$$

such that $\tilde{p}\ ([d]^r) = [\tilde{p}\ (\tau)|\tau \in [d]^r\ ]$. Thus, $[\widehat{Z}]^r$ can be determined as follows

$$[\widehat{Z}]^r = \left(\underline{\widehat{Z}}^r, \overline{\widehat{Z}}^r\right) = ((\underline{d}^r, \underline{P}^r), \left(\overline{d}^r, \overline{P}^r\right)) \tag{4.8}$$

such that $\underline{P}^r = \underline{d}^r\ \tilde{p}\ \left(\underline{\tau}_j^r\right), \overline{P}^r = \overline{d}^r\ \tilde{p}\ (\overline{\tau}_j^\alpha), [\tau_j]^r = (\underline{\tau}_j^r, \overline{\tau}_j^\alpha)$.

Similar with fuzzy numbers[21–24], some operations such as sum, subtract and multiply denoted by $\oplus$, $\ominus$ and $\odot$, respectively can be also defined for $\widehat{Z}$ -numbers.

Assume that $\widehat{Z}_1 = (d_1, \tilde{p}_1)$ and $\widehat{Z}_2 = (d_2, \tilde{p}_2)$ are two discrete $\widehat{Z}$ -numbers describing the uncertain variables τ_1 and τ_2 thus, $\sum_{j=1}^{n} \tilde{p}_1(\tau_{1j}) = 1, \sum_{j=1}^{n} \tilde{p}_1(\tau_{2j}) =$

1. For $\widehat{Z}_1 = (d_1, \widetilde{p}_1)$ and $\widehat{Z}_2 = (d_2, \widetilde{p}_2)$ the below-mentioned operation can be defined

$$\widehat{Z}_{12} = \widehat{Z}_1 * \widehat{Z}_2 = (d_1 * d_2, \widetilde{p}_1 * \widetilde{p}_2) \tag{4.9}$$

such that $* \in (\oplus, \ominus, \odot)$.

The operations implemented for the fuzzy numbers $[d_1]^r = [d_{11}^r, d_{12}^r]$ and $[d_2]^r = [d_{21}^r, d_{22}^r]$ can be determined as [21, 25, 26],

$$\begin{aligned} [d_1 \oplus d_2]^r &= [d_1]^r + [d_2]^r = [d_{11}^r + d_{21}^r, d_{12}^r + d_{22}^r] \\ [d_1 \ominus d_2]^r &= [d_1]^r - [d_2]^r = [d_{11}^r - d_{22}^r, d_{12}^r - d_{21}^r] \\ [d_1 \odot d_2]^r &= \begin{pmatrix} \min[d_{11}^r d_{21}^r, d_{11}^r d_{22}^r, d_{12}^r d_{21}^r, d_{12}^r d_2^r] \\ \max[d_{11}^r d_{21}^r, d_{11}^r d_{22}^r, d_{12}^r d_{21}^r, d_{12}^r d_2^r] \end{pmatrix} \end{aligned} \tag{4.10}$$

For all $\widetilde{p}_1 * \widetilde{p}_2$ operations the below-mentioned equation can be given considering the discrete probability distributions,

$$\widetilde{p}_1 * \widetilde{p}_2 = \sum_{\iota} \widetilde{p}_1(\tau_{1,j}) \widetilde{p}_2(\tau_{2,(n-j)}) = \widetilde{p}_{12}(\tau) \tag{4.11}$$

Definition 4 (*Hukuhara difference*) The Hukuhara difference can be stated as follows [27],

$$\begin{aligned} \widehat{Z}_1 \ominus_H \widehat{Z}_2 &= \widehat{Z}_{12} \\ \widehat{Z}_1 &= \widehat{Z}_2 \oplus \widehat{Z}_{12} \end{aligned} \tag{4.12}$$

If $\widehat{Z}_1 \ominus_H \widehat{Z}_2$ exists, the r -level can be stated as follows

$$[\widehat{Z}_1 \ominus_H \widehat{Z}_2]^r = [\underline{Z}_1^r - \underline{Z}_2^r, \overline{Z}_1^r - \overline{Z}_2^r] \tag{4.13}$$

It is clear that, $\widehat{Z}_1 \ominus_H \widehat{Z}_1 = 0$, $\widehat{Z}_1 \ominus \widehat{Z}_1 \neq 0$

Additionally, the generalized Hukuhara difference can be stated as follows [28],

$$\widehat{Z}_1 \ominus_{gH} \widehat{Z}_2 = \widehat{Z}_{12} \Leftrightarrow \begin{cases} 1) \widehat{Z}_1 = \widehat{Z}_2 \oplus \widehat{Z}_{12} \\ 2) \widehat{Z}_2 = \widehat{Z}_1 \oplus (-1) \widehat{Z}_{12} \end{cases} \tag{4.14}$$

Applying the r-level to (4.14), we have $[\widehat{Z}_1 \ominus_{gH} \widehat{Z}_1]^r = [\min\left(\underline{Z}_1^r - \underline{Z}_2^r, \overline{Z}_1^r - \overline{Z}_2^r\right), \max\left(\underline{Z}_1^r - \underline{Z}_2^r, \overline{Z}_1^r - \overline{Z}_2^r\right)$ [29–34] also in a case $\widehat{Z}_1 \ominus_{gH} \widehat{Z}_2$ and $\widehat{Z}_1 \ominus_H \widehat{Z}_2$ exists $\widehat{Z}_1 \ominus_H \widehat{Z}_2 = \widehat{Z}_1 \ominus_{gH} \widehat{Z}_2$. The following conditions are defined for the existence of $\widehat{Z}_{12} = \widehat{Z}_1 \ominus_{gH} \widehat{Z}_{12} \in E$

$$1)\begin{cases} \widehat{\underline{Z}}_{12}^{r} = \widehat{\underline{Z}}_{1}^{r} - \widehat{\underline{Z}}_{2}^{r} and \overline{\widehat{Z}}_{12}^{r} = \overline{\widehat{Z}}_{1}^{r} - \overline{\widehat{Z}}_{2}^{r} \\ with \widehat{\underline{Z}}_{12}^{r} increasing, \overline{\widehat{Z}}_{12}^{r} decrease, \widehat{\underline{Z}}_{12}^{r} \leq \overline{\widehat{Z}}_{12}^{r} \end{cases}$$
$$1)\begin{cases} \widehat{\underline{Z}}_{12}^{r} = \overline{\widehat{Z}}_{1}^{r} - \overline{\widehat{Z}}_{2}^{r} and \overline{\widehat{Z}}_{12}^{r} = \widehat{\underline{Z}}_{1}^{r} - \widehat{\underline{Z}}_{2}^{r} \\ with \widehat{\underline{Z}}_{12}^{r} increasing, \overline{\widehat{Z}}_{12}^{r} decrease, \widehat{\underline{Z}}_{12}^{r} \leq \overline{\widehat{Z}}_{12}^{r} \end{cases} \tag{4.15}$$

such that $\forall r \in [0, 1]$.

Assume that d is a triangular function, the absolute value of the $\widehat{Z}$ -number $\widehat{Z} = (d, \widetilde{p})$ can be stated as follows

$$\left|\widehat{Z}(\tau)\right| = (|h_1| + |k_1| + |l_1|, \widetilde{p}\ (|h_2| + |k_2| + |l_2|)) \tag{4.16}$$

Let us assume that d_1 and d_2 are triangular functions, the supremum metric for $\widehat{Z}$-numbers $\widehat{Z}_1 = (d_1, \widetilde{p}_1)$ and $\widehat{Z}_2 = (d_2, \widetilde{p}_2)$ can be stated as below

$$D\left(\hat{Z}_1, \hat{Z}_2\right) = d(d_1, d_2) + d(\tilde{p}_1, \tilde{p}_2) \tag{4.17}$$

such that $d(.,.)$ is taken to be the supremum metrics for fuzzy sets [26]. The following properties can be defined for $D\left(\hat{Z}_1, \hat{Z}_2\right)$

$$\begin{aligned} D\left(\hat{Z}_1 + \hat{Z}, \hat{Z}_2 + \hat{Z}\right) &= D\left(\hat{Z}_1, \hat{Z}_2\right) \\ D\left(\hat{Z}_2, \hat{Z}_1\right) &= D\left(\hat{Z}_1, \hat{Z}_2\right) \\ D\left(c\hat{Z}_2, c\hat{Z}_1\right) &= |c| D\left(\hat{Z}_1, \hat{Z}_2\right) \\ D\left(\hat{Z}_1, \hat{Z}_2\right) &\leq D\left(\hat{Z}_1, \hat{Z}\right) + D\left(\hat{Z}, \hat{Z}_2\right) \end{aligned} \tag{4.18}$$

such that $c \in \Re$, $\widehat{Z} = (d, \widetilde{p})$ is taken to be $\widehat{Z}$ -number and d is taken to be triangle function.

Assume that $\overline{Z}$ is the space of $\hat{Z}$ -numbers, hence the r- level of the $\hat{Z}$ -number valued function $Q : [0, h] \rightarrow \overline{Z}$ can be stated as below

$$Q(d, r) = \left[\underline{Q}(d, r), \overline{Q}(d, r)\right] \tag{4.19}$$

such that $d \in \overline{Z}$, for each $r \in [0, 1]$.

By implementing the definition of Generalized Hukuhara difference, the gH-derivative of Q at d_0 can be expressed as below

$$\frac{d}{dt} Q(d_0) = \lim_{\rho \to 0} \frac{1}{\rho}\left[Q(d_0 + \rho) \oplus_{gH} Q(d_0)\right] \tag{4.20}$$

In (4.20), $Q(d_0 + \rho)$ and $Q(d_0)$ demonstrate symmetric pattern with $\hat{Z}_1$ and $\hat{Z}_2$ respectively defined in (4.14).

Definition 5 (*distance*) The distance between two $\hat{Z}$ -numbers, $u, v \in \overline{Z}$ can be stated as the Hausdorff metric $\tilde{D}_Q[u, v]$,

$$\tilde{D}_Q[u, v] = \max[sup_{(b_1, d_1 \in u)} inf_{(b_2, d_2 \in v)} (\tilde{d}(b_1, b_2) + \tilde{d}(d_1, d_2), sup_{(b_1, d_1 \in u)} inf_{(b_2, d_2 \in v)} \left(\tilde{d}(b_1, b_2) + \tilde{d}(b_1, d_2)\right] \quad (4.21)$$

$\tilde{d}(b, d)$ is taken to be the supremum metrics described for fuzzy sets.

Lemma 1 *Assume that is a compact set, hence, $\zeta \subset \overline{Z}$ is uniformly support-bounded, such that there is a compact set $H \subset \Re$, such that $\forall u \in \zeta$,*

$$Supp(u) \subset H \quad (4.22)$$

Lemma 2 *Let $u, v \in \overline{Z}$ and $r \in (0, 1]$, hence we have: (i) if $\varphi : \Re \to \Re$ is continuous, $[\varphi(u)]^r = \varphi([u^r])$ holds; (ii) if $\varphi : \Re \to \Re$ is continuous, hence $\varphi(Supp(u)) = Supp(\varphi(u))$.*

Proof Whereas (i) concludes from [35], therefore only (ii) will be proved. At the start, $\overline{\varphi(B_1)} = \varphi(\overline{B_1})$ is proved for $B_1 \subset \Re$. Whereas $\varphi(B_1) \subset \varphi(\overline{B_1})$, furthermore $\varphi(\overline{B_1})$ is closed because of the continuity of φ, thus, $\underline{\varphi(B_1)} \subset \varphi(\overline{B_1})$.

In addition, for randomly given $\varpi = \varphi(\overline{B_1})$, there is a sequence $\{\gamma_\iota | \iota \in N\} \subset \Re$, $\gamma \in \Re$, such that $\gamma_\iota \to \gamma(\iota \to +\infty)$, $\varpi = \varphi(\gamma)$. Whereas φ is continues hence, $\lim_{\iota \to +\infty} \varphi(\gamma_\iota) = \varphi(\gamma) = \varpi$. However, $\varphi(\gamma) \in \varphi(B_1)$, thus, $\varpi \in \overline{\varphi(B_1)}$. Hence $\varphi(\overline{B_1}) \subset \overline{\varphi(B_1)}$. Accordingly,$\overline{\varphi(B_1)} = \varphi(\overline{B_1})$.

So,

$$Supp(\varphi(u)) = \overline{\varpi \in \Re | (\varphi(u))(\varpi) > 0)}$$
$$\varphi(Supp(u)) = \varphi\left(\overline{\gamma \in \Re | u(\gamma) > 0}\right) \quad (4.23)$$

Thus

$$\varphi(Supp(u)) = \overline{\varphi(\gamma \in \Re | u(\gamma) > 0)} = (\varphi(\gamma) \in \Re | u(\gamma) > 0) \quad (4.24)$$

Since, $[\varpi \in \Re | (\varphi(u))(\varpi)] > 0 = [\varphi(\gamma) | u(\gamma) > 0]$, thus

$$Supp(\varphi(u)) = \varphi(Supp(u)) \quad (4.25)$$

Lemma 3 *Assume that $\Gamma \subset \Re$ is a compact set, furthermore φ_1 and φ_2 are taken to be continuous on Γ, $\kappa > 0$, also.*

$$|\varphi_1(c) - \varphi_2(c)| < \kappa, \forall c \in \Gamma \tag{4.26}$$

Thus $\left|sup_{c\in\Gamma_1}\varphi_1(c) - sup_{c\in\Gamma_1}\varphi_2(c)\right| < \kappa$ satisfies for each compact set $\Gamma_1 \subset \Gamma$.

Theorem 1 *Let $\varphi : \Re \to \Re$ is a continuous function, furthermore for each compact set $b \subset \overline{Z}_0$ (the set of all the bounded $\hat{Z}$ -number set) moreover $\phi > 0$, there is $c_j \in \overline{Z}_0$, $j = 1, 2, \ldots, m$, such that*

$$\tilde{d}\left(\varphi(\alpha), \sum_{j=1}^{m} \varphi_j(\beta)c_j\right) < \phi \quad \forall\alpha \in b, \ \forall\beta \in \Re \tag{4.27}$$

such that ϕ is taken to be a finite number.

Theorem 2 *If $\varphi : \Re \to \Re$ is a continuous function, hence for each compact set $b \subset \overline{Z}_0$, $x > 0$ and arbitrary $\epsilon > 0$, there is $c_j \in \overline{Z}_0$, $j = 1, 2, \ldots, m$, such that*

$$\tilde{d}\left(\varphi(\alpha), \sum_{j=1}^{m} \varphi_j(\alpha)c_j\right) < x, \forall\alpha \in b \tag{4.28}$$

such that, X is taken to be a finite number. The lower and the upper limits of the r-level set of $\hat{Z}$ -number function go towards X. Nevertheless, the center goes towards ϵ.

Theorem 3 *Assume that $b \subset \overline{Z}_0$ is compact, moreover H is taken to be the corresponding compact set of b also $\varphi, \hat{\varphi} : \Re \to \Re$ are taken to be the continuous functions such that,*

$$\left|\varphi(\alpha) - \hat{\varphi}(\alpha)\right| < \kappa, \forall\alpha \in H \tag{4.29}$$

where $\kappa > 0$. So,$\forall\eta \in b, \tilde{d}\big(\varphi(\eta) - \hat{\varphi}(\eta)\big) \le \kappa$.

4.2 Fuzzy Logic Systems

Fuzzy logic systems have been widely developed and used in many areas of industries to be mentioned as automobile speed control [36], robot arm control [37], water quality control [38], and automatic train operation systems [39]. They can be effectively applied for increasing the productivity of the system. Fuzzy logic system is an extremely appropriate tool for estimated reasoning, especially for the system with a mathematical model which is quite difficult to achieve [23–26, 34, 40–42].

Fuzzy If–Then Rules
Fuzzy logic commonly utilizes If–Then rules [43–47]. In general, a single fuzzy If–Then rule can be represented as follows:

$$If u is f_1 Then\ v\ is\ g_2$$

in which f_1 and g_2 are considered as the linguistic variables defined by fuzzy sets on the domains U and V, respectively.

The If sector of the rule "u is f_1" is termed as the antecedent also the Then sector of the rule "v is g_2" is termed as the consequent.

Moreover, the conditional statement can be given as follows:$If\ f_1\ Then\ g_2\ or\ f_1 \rightarrow g_2$

4.2.1 Example 1

Consider a steam engine such that the speed and the pressure of that is defined in the form of the following linguistic expression

$$If\ speed\ is\ slow\ then\ pressure\ is\ high$$

This linguistic expression can be demonstrated as Fig. 4.1.

Fuzzy expert system The fuzzy expert system uses fuzzy membership functions and rules, in place of the Boolean logic [48–52]. Fuzzy expert system is considered as the most common utilization of fuzzy logic. It can be used in several areas such as:

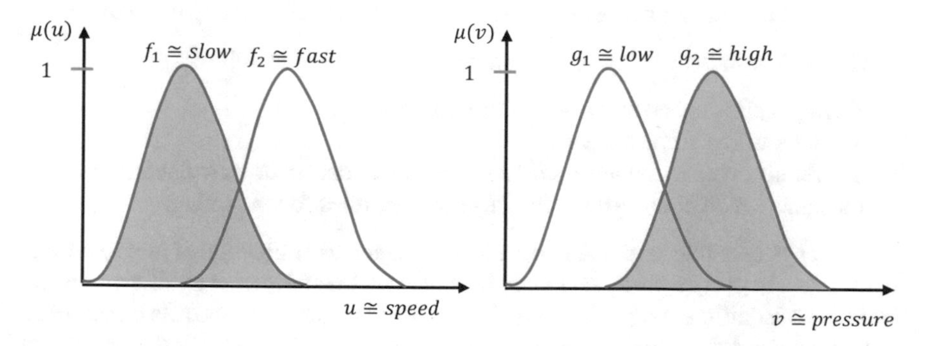

Fig. 4.1 Fuzzy If–Then rule

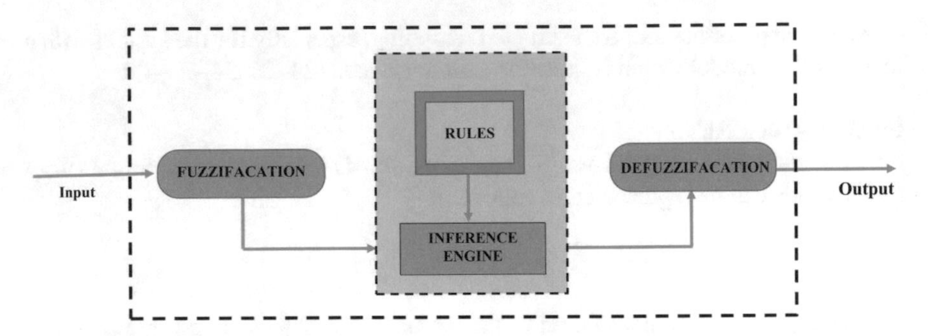

Fig. 4.2 The framework of the fuzzy expert system

- Linear and nonlinear control
- Automatic transmission control
- Robots decision-making

The fuzzy inference systems for industrial position factor analysis are made up of three sections named the fuzzifier, the knowledge management and the defuzzifier (Fig. 4.2). The two sections, fuzzifier and the defuzzifier are used for transforming exterior information in fuzzy values and vice versa. The knowledge management section is used to apply the fuzzy rules concerned with the knowledge basis and permits for estimate logic [53].

Beneficial and Weak Points of Fuzzy Based Systems

The beneficial of fuzzy based systems can be named as follows:

1. Results with perfect accuracy.
2. Rational as human.
3. Implementing basic mathematical models for solving real world problems concerning linear or non-linear systems.
4. Implemented for quick operations as well as decisions regarding the control.
5. Highly efficient for rule-based modeling as well as membership evaluations.

The weak points of fuzzy based systems can be named as follows:

1. Having a slow speed and also requires more time to run.
2. The lack of real-time answer.
3. The requirement to involve a sufficient amount of data in order to have an accurate outcome which could lead to the increment of rules for reasoning.

Fuzzy systems find increasing use in detection and identification of multiple leaks in pipelines. In [54] the fuzzy set theory is applied successfully to an intelligent system for leakage identification in pipelines. The obtained results from the rule based fuzzy system demonstrates that the fuzzy logic implemented for evaluating a petroleum derivate transference procedure is an efficient and useful tool.

In [55] an efficient diagnosis and localization scheme based on fuzzy decision-making for the leakage detection in oil pipelines is presented. Two sensors required for that technique have been installed at each terminal of pipeline. The technique accurately determines the exact location of the leak.

Examples This section provides several real examples in order to demonstrate the application of fuzzy logic in control of flowing fluids.

4.2.2 Example 2

The cold-water storage cistern is made of two inlet pipes D_1, D_2 and two exhaust pipes D_3, D_4. The cross-sectional areas associated with pipes are $K_1 = Q(0.020, 0.021, 0.023)$, $K_2 = Q(0.014, 0.024, 0.032)$, $K_3 = Q(0.002, 0.006, 0.008) K_4 = Q(0.039, 0.049, 0.058)$. The flow speed in pipes D_1,D_2, D_3, and D_4 are $s_1 = \left(\frac{a}{3}\right)a$, $s_2 = e^a \sin(a)$, $s_3 = \sin\left(\frac{a}{2}\right)$, $s_4 = \frac{a}{3}$, respectively. The control variable should be obtained in such a way that $c = (6.522, 9.899, 39.981)$. The general formula for a mass balance associated with the cold-water storage cistern is defined as follows [56],

$$\varsigma K_1 s_1 \oplus \varsigma K_2 s_2 = \varsigma K_3 s_3 \oplus \varsigma K_4 s_4 \oplus \varsigma \tag{4.30}$$

in which ς is taken to be the density of the water and also the exact solution is take to be $a_0 = 2.5$. The estimated solution of (30) can be obtained using Steepest descent approach [57], Adomian decomposition approach [58], and Ranking approach [59]. The errors associated with these approaches are represented in Fig. 4.3.

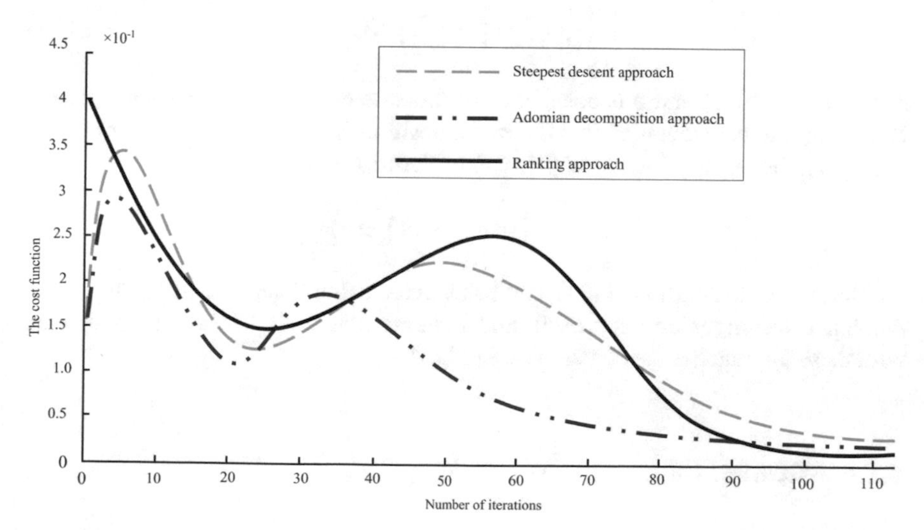

Fig. 4.3 The errors associated with three common approaches

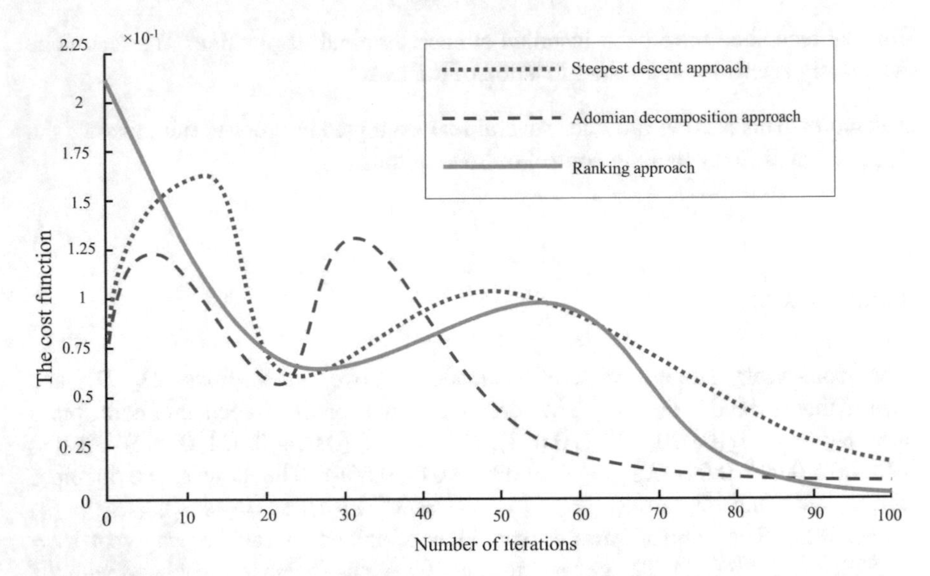

Fig. 4.4 The errors associated with three common approaches

4.2.3 Example 3

To develop a mathematical model of a heating system inside a water tank we suppose that the thermal capacitance is $M = 2.5$, resistance is $W = 1$ and the temperature is υ. The mathematical modeling can be described as follows [60],

$$\frac{d}{dt}\upsilon(t) = -\frac{1}{WM}\upsilon(t) \tag{4.31}$$

such that $t \in [0, 1]$ and υ is considered as the amount of sinking in each moment. By taking into consideration that the initial position is $\upsilon(0) = (r - 1, 1 - r)$ where $r \in [0, 1]$, the exact solutions of (31) can be described as follows:

$$\upsilon(t, r) = \left[(r - 1)e^t, (1 - r)e^t\right] \tag{4.32}$$

The estimated solution of (31), can be obtained using Steepest descent approach, Adomian decomposition approach, and Ranking approach. The errors associated with these approaches are represented in Fig. 4.4.

4.2.4 Example 4

The inflow of fluid into a tank system generates disturbance within the system and produces vibration in liquid level U and is formulated as $A = t + 1$. Assume that the flow obstruction is $S = 1$ and the cross section associated with the tank is $G = 1$. The mathematical modeling of the liquid level can be described as follows [56],

$$\begin{cases} U'(t) = -\frac{1}{GS}U(t) + \frac{A}{G}, 0 \leq t \leq T \\ U(0) = \left[\left(\underline{U}(0,r), \overline{U}(0,r)\right), p(0.8, 0.9, 1)\right] \end{cases} \tag{4.33}$$

The following relation can be obtained applying the fuzzy Sumudu transform approach based on Z-number

$$-S[U(t)] = \left(\frac{1}{C} \odot S[U(t)]\right) \ominus \left(\frac{1}{C} S[U(0)]\right) \tag{4.34}$$

$$S\left[U'(t)\right]\left(\int_a^{\infty} U'(Ct)e^{-t}dt\right) \tag{4.35}$$

where $0 \leq C \leq J$, $J \geq 0$. We have

$$S\left[U'(t)\right] = \left(\frac{-1}{C} \odot S[U(t)]\right) \ominus \left(\frac{-1}{C} S[U(0)]\right) \tag{4.36}$$

Hence,

$$-S[U(t)] + S[t] + S[1] = \left(\frac{-1}{C} \odot S[U(t)]\right) \ominus \left(\frac{-1}{C} S[U(0)]\right) \tag{4.37}$$

So,

$$\begin{cases} -S\left[\underline{U}(t)\right] + S[t] + S[1] = \frac{1}{C} S\left[\underline{U}(t,r)\right] - \frac{1}{C}\underline{U}(0,r) \\ -S\left[\overline{U}(t)\right] + S[t] + S[1] = \frac{1}{C} S\left[\overline{U}(t,r)\right] - \frac{1}{C}\overline{U}(0,r) \end{cases} \tag{4.38}$$

The Z-number solution of Eq. (4.38) is acquired in the following form,

$$\begin{cases} S\left[\underline{U}(t,r)\right] = S[t] + S[1] + \frac{-1}{C} S\left[\underline{U}(t,r)\right] - \frac{1}{C}\underline{U}(0,r) \\ S\left[\overline{U}(t,r)\right] = S[t] + S[1] + \frac{1}{C} S\left[\overline{U}(t,r)\right] - \frac{1}{C}\overline{U}(0,r) \end{cases} \tag{4.39}$$

Thus,

$$\begin{cases} S\left[\underline{U}(t,r)\right] = \left(\frac{1}{C+1}\right)\underline{U}(0,r) + C \\ S\left[\overline{U}(t,r)\right] = \left(\frac{1}{C+1}\right)\overline{U}(t,r) + C \end{cases} \tag{4.40}$$

By implementing the inverse Sumudu transform for Z-numbers, the following relation is acquired

$$\begin{cases} \underline{U}(t,r) = \underline{U}(0,r)S^{-1}\left(\frac{1}{C+1}\right) + S^{-1}(C) \\ \overline{U}(t,r) = \overline{U}(t,r)S^{-1}\left(\frac{1}{C+1}\right) + S^{-1}(C) \end{cases} \tag{4.41}$$

such that

$$\begin{cases} \underline{U}(t,r) = e^{-t}\underline{U}(0,r) + t \\ \overline{U}(t,r) = e^{-t}\overline{U}(t,r) + t \end{cases} \tag{4.42}$$

In a case that the initial condition is in the form of Z-number, $U(0) = [(-a(1-r), a(1-r)), p(0.75, 0.8, 0.95)]$, we have

$$\begin{cases} \underline{U}(t,r) = e^{-t}(-a(1-r)) + t \\ \overline{U}(t,r) = e^{-t}(a(1-r)) + t \end{cases} \tag{4.43}$$

Table 4.1 displays the estimated errors between exact solutions and approximate solutions for Z -numbers using the fuzzy Sumudu transform approach. Figure 4.5 displays the exact solution plot for fuzzy numbers.

Table 4.1 The estimated errors between exact solutions and approximate solutions for Z-numbers

r	The estimated errors between exact solutions and approximate solutions
0	[(0.0062, 0.0141), p(0.7, 0.8, 0.91)]
0.2	[(0.0022, 0.0135), p(0.82, 0.94, 1)
0.4	[(0.0043, 0.0092), p(0.75, 0.8, 0.92)]
0.6	[(0.0009, 0.0033), p (0.75, 0.8, 0.93)
0.8	[(0.0059, 0.0093), p(0.85, 0.96,1)
1	[(0.0122, 0.0122), p (0.86, 0.91, 1)

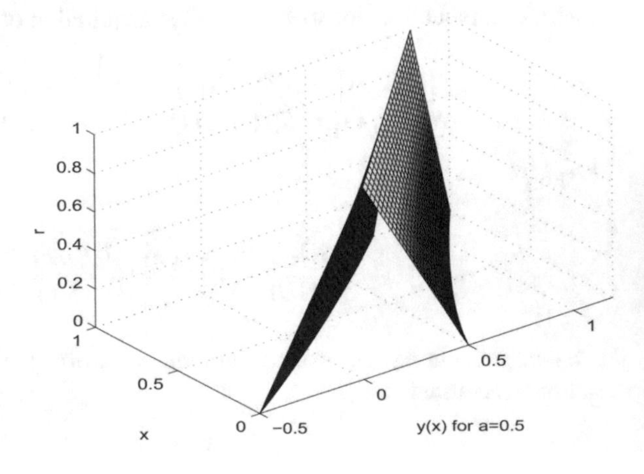

Fig. 4.5 The exact solution plot for fuzzy numbers

4.3 Conclusions

In this chapter, in order to control of flowing fluids the concept of fuzzy logic has been studied. One of the principal contributions of fuzzy logic is its capability to provide a basis to deal with vague information. Fuzzy logic can be efficiently used in designing systems which cannot define exactly as it does not need an accurate mathematical model. In this chapter, we have presented the uncertainty in the form of fuzzy numbers and Z-numbers. Several real examples are implemented to demonstrate the application of fuzzy logic in control of flowing fluids.

References

1. Jafari, R., Yu, W., Li, X.: Fuzzy differential equations for nonlinear system modeling with Bernstein neural networks. IEEE Access **4**, 9428–9436 (2016)
2. Jafari, R., Yu, W.: Fuzzy modeling for uncertainty nonlinear systems with fuzzy equations. Math. Probl. Eng. (2017)
3. Jafari, R., Yu, W., Razvarz, S., Gegov, A.: Numerical methods for solving fuzzy equations: a survey. Fuzzy Sets Syst. (2019)
4. Jafarian, A., Jafari, R.: A new computational method for solving fully fuzzy nonlinear matrix equations. Int. J. Fuzzy Comput. Modell. **2**(4), 275–285 (2019)
5. Jafarian, A., Jafari, R.: Simulation and evaluation of fuzzy polynomials by feed-back neural networks (2012)
6. Jafarian, A., Jafari, R.: Approximate solutions of dual fuzzy polynomials by feed-back neural networks. J. Soft Comput. Appl. **2012**, 1–5 (2012)
7. Jiang, W., Xie, C., Luo, Y., Tang, Y.: Ranking Z-numbers with an improved ranking method for generalized fuzzy numbers. J. Intell. Fuzzy Syst. **32**(3), 1931–1943 (2017)
8. Zamri, N., Ahmad, F., Rose, A.N.M., Makhtar, M.: A fuzzy TOPSIS with Z-numbers approach for evaluation on accident at the construction site. In: International Conference on Soft Computing and Data Mining, , pp 41–50. Springer (2016)
9. Abiyev, R.H., Uyar, K., Ilhan, U., Imanov, E., Abiyeva, E.: Estimation of food security risk level using Z-number-based fuzzy system. J. Food Qual. (2018)
10. Aliev, R.A., Alizadeh, A.V., Huseynov, O.H.: The arithmetic of discrete Z-numbers. Inf. Sci. **290**, 134–155 (2015)
11. Aliev, R.A., Huseynov, O.H., Zeinalova, L.M.: The arithmetic of continuous Z-numbers. Inf. Sci. **373**, 441–460 (2016)
12. Babanli, M., Huseynov, V.: Z-number-based alloy selection problem. Procedia Comput. Sci. **102**, 183–189 (2016)
13. Jafari, R., Razvarz, S., Gegov, A.: Neural network approach to solving fuzzy nonlinear equations using Z-numbers. IEEE Trans. Fuzzy Syst. (2019)
14. Jafari, R., Yu, W., Li, X.: Numerical solution of fuzzy equations with Z-numbers using neural networks. Intell. Autom. Soft Comput., 1–7 (2017)
15. Jafari, R., Yu, W., Li, X., Razvarz, S.: Numerical solution of fuzzy differential equations with Z-numbers using Bernstein neural networks. Int. J. Comput. Intell. Syst. **10**(1), 1226–1237 (2017)
16. Jafari, R., Razvarz, S., Gegov, A., Paul, S.: Fuzzy modeling for uncertain nonlinear systems using fuzzy equations and Z-numbers. In: UK Workshop on Computational Intelligence, pp. 96–107. Springer (2018)
17. Jafari, R., Yu, W.: Uncertain nonlinear system control with fuzzy differential equations and Z-numbers. In: 2017 IEEE International Conference on Industrial Technology (ICIT), pp. 890–895. IEEE (2017)

18. Jafari, R., Razvarz, S., Gegov, A.: Paul S Modeling and control of uncertain nonlinear systems. In: 2018 International Conference on Intelligent Systems (IS), pp. 168–173. IEEE (2018)
19. Jafari, R., Razvarz, S., Gegov, A., Yu, W.: Fuzzy control of uncertain nonlinear systems with numerical techniques: a survey. In: UK Workshop on Computational Intelligence, , pp. 3–14. Springer (2019)
20. Jafari, R., Yu, W.: Uncertainty nonlinear systems modeling with fuzzy equations. In: 2015 IEEE International Conference on Information Reuse and Integration, pp. 182–188. IEEE (2015)
21. de Barros, L.C., Bassanezi, R.C., Lodwick, W.A.: The extension principle of Zadeh and fuzzy numbers. In: de Barros LC, Bassanezi RC, Lodwick WA (eds) A First Course in Fuzzy Logic, Fuzzy Dynamical Systems, and Biomathematics: Theory and Applications. Springer Berlin Heidelberg, Berlin, Heidelberg, pp 23–41 (2017). https://doi.org/10.1007/978-3-662-53324-6_2
22. Jafari, R., Yu, W.: Uncertainty nonlinear systems control with fuzzy equations. In: 2015 IEEE International Conference on Systems, Man, and Cybernetics, pp 2885–2890. IEEE (2015)
23. Jafari, R., Yu, W., Li, X.: Solving fuzzy differential equation with Bernstein neural networks. In: 2016 IEEE International Conference on Systems, Man, and Cybernetics (SMC), pp. 001245–001250. IEEE (2016)
24. Jafarian, A., Jafari, R.: An iterative method for solving fuzzy polynomials by fuzzy neural networks (2012)
25. Jafarian, A., Jafari, R., Al Qurashi, M.M., Baleanu, D.: A novel computational approach to approximate fuzzy interpolation polynomials. SpringerPlus **5**(1), 1428 (2016)
26. Jafari, R., Yu, W.: Fuzzy control for uncertainty nonlinear systems with dual fuzzy equations. J. Intell. Fuzzy Syst. **29**(3), 1229–1240 (2015)
27. Aliev, R.A., Pedrycz, W., Kreinovich, V., Huseynov, O.H.: The general theory of decisions. Inf. Sci. **327**, 125–148 (2016)
28. Bede, B., Stefanini, L.: Generalized differentiability of fuzzy-valued functions. Fuzzy Sets Syst. **230**(1), 119–141 (2013)
29. Jafari, R., Razvarz, S., Gegov, A.: A novel technique to solve fully fuzzy nonlinear matrix equations. In: International Conference on Theory and Applications of Fuzzy Systems and Soft Computing, pp. 886–892. Springer (2018)
30. Jafari, R., Razvarz, S., Gegov, A., Paul, S., Keshtkar, S.: Fuzzy Sumudu transform approach to solving fuzzy differential equations with Z-numbers. In: Advanced Fuzzy Logic Approaches in Engineering Science, pp. 18–48. IGI Global (2019)
31. Jafarian, A., Jafari, R.: New iterative approach for solving fully fuzzy polynomials
32. Jafarian, A., Jafari, R., Golmankhaneh, A.K., Baleanu, D.: Solving fully fuzzy polynomials using feed-back neural networks. Int. J. Comput. Math. **92**(4), 742–755 (2015)
33. Razvarz, S., Jafari, R., Yu, W.: Numerical solution of fuzzy differential equations with Z-numbers using fuzzy Sumudu transforms. Adv. Sci. Technol. Eng. Syst. J. (ASTESJ) **3**, 66–75 (2018)
34. Jafari, R., Razvarz, S., Gegov, A.: Fuzzy differential equations for modeling and control of fuzzy systems. In: International Conference on Theory and Applications of Fuzzy Systems and Soft Computing, pp 732–740. Springer (2018)
35. Yang, L., Gao, Y.: Fuzzy Mathematics-Theory and its Application. South China University of Technology, Guangzhou (1993)
36. Murakami, S., Maeda, M.: Automobile speed control system using a fuzzy logic controller. Indus. Appl. Fuzzy Control (1985)
37. Scharf, E.: The application of a fuzzy controller to the control of a multi-degree-freedom robot arm. Industrial applications of fuzzy control (1985)
38. Yagishita, O.: Application of fuzzy reasoning to the water purification process. Indus. Appl. Fuzzy Control, 19–40 (1985)
39. Yasunobu, S.: Automatic train operation by predictive fuzzy control. Indus. Appl. Fuzzy Control, 1–18 (1985)
40. Jafari, R., Razvarz, S.: Solution of fuzzy differential equations using fuzzy Sumudu transforms. In: Optimization in Control Applications, vol 23. Mathematical and Computational Applications, pp. 186–200 (2018)

41. Jafari, R., Razvarz, S.: Solution of fuzzy differential equations using fuzzy Sumudu transforms. Math. Comput. Appl. **23**(1), 5 (2018)
42. Jafari, R., Razvarz, S., Gegov, A.: Solving differential equations with Z-numbers by utilizing fuzzy sumudu transform. In: Proceedings of SAI Intelligent Systems Conference, pp. 1125–1138. Springer (2018)
43. Sugeno, M.: An introductory survey of fuzzy control. Inf. Sci. **36**(1–2), 59–83 (1985)
44. Wang, L.X., Mendel, J.M.: Generating fuzzy rules by learning from examples. IEEE Trans. Syst. Man Cybern. **22**(6), 1414–1427 (1992)
45. Takagi, T., Sugeno, M.: Fuzzy identification of systems and its applications to modeling and control. IEEE Trans. Syst. Man Cybern. **1**, 116–132 (1985)
46. Lee, C-C.: Fuzzy logic in control systems: fuzzy logic controller. I. IEEE Trans. Syst. Man Cybern. **20**(2), 404–418 (1990)
47. Sugeno, M., Yasukawa, T.: A fuzzy-logic-based approach to qualitative modeling. IEEE Trans. Fuzzy Syst. **1**(1), 7–31 (1993)
48. Dash, P., Liew, A., Rahman, S., Ramakrishna, G.: Building a fuzzy expert system for electric load forecasting using a hybrid neural network. Expert Syst. Appl. **9**(3), 407–421 (1995)
49. Kahramanli, H., Allahverdi, N.: Design of a hybrid system for the diabetes and heart diseases. Expert Syst. Appl. **35**(1–2), 82–89 (2008)
50. Polat, K., Güneş, S., Arslan, A.: A cascade learning system for classification of diabetes disease: Generalized discriminant analysis and least square support vector machine. Expert Syst. Appl. **34**(1), 482–487 (2008)
51. Polat, K., Güneş, S.: An expert system approach based on principal component analysis and adaptive neuro-fuzzy inference system to diagnosis of diabetes disease. Dig. Sig. Process. **17**(4), 702–710 (2007)
52. Buckley, J., Siler, W., Tucker, D.: A fuzzy expert system. Fuzzy Sets Syst. **20**(1), 1–16 (1986)
53. Czogala, E., Leski, J.: Fuzzy and neuro-fuzzy intelligent systems, vol 47. Physica (2012)
54. Da Silva, H.V., Morooka, C.K., Guilherme, I.R., da Fonseca, T.C., Mendes, J.R.: Leak detection in petroleum pipelines using a fuzzy system. J. Petrol. Sci. Eng. **49**(3–4), 223–238 (2005)
55. Feng, J., Zhang, H., Liu, D.: Applications of fuzzy decision-making in pipeline leak localization. In: 2004 IEEE International Conference on Fuzzy Systems (IEEE Cat. No. 04CH37542), , pp. 599–603. IEEE (2004)
56. Wylie, E.B., Streeter, V.L.: Fluid transients. mhi (1978)
57. Abbasbandy, S., Jafarian, A.: Steepest descent method for solving fuzzy nonlinear equations. Appl. Math. Comput. **174**(1), 669–675 (2006)
58. Paripour, M., Hajilou, E., Hajilou, A., Heidari, H.: Application of Adomian decomposition method to solve hybrid fuzzy differential equations. J. Taibah Univ. Sci. **9**(1), 95–103 (2015)
59. Delgado, M., Vila, M.A., Voxman, W.: On a canonical representation of fuzzy numbers. Fuzzy Sets Syst. **93**(1), 125–135 (1998)
60. Anderson, D., Tannehill, J.C., Pletcher, R.H.: Computational Fluid Mechanics and Heat Transfer. Taylor & Francis (2016)

Chapter 5
Basic Concepts of Neural Networks and Deep Learning and Their Applications for Pipeline Damage Detection

5.1 Different Types of Threats Occurring in Pipeline Systems

In this section different types of threats occurring in pipeline systems are presented.

Metal loss in pipe Metal loss defects happen due to the reduction of wall thickness of the pipes caused by internal or external corrosion processes. Metal loss defects in pipeline is displayed in Fig. 5.1.

Dent in pipe Dent is a common form of defect in the pipe and is a permanent plastic metamorphosis of the circular cross, see Fig. 5.2. A dent causes the pipe's diameter to decrease and leads to strain concentrations.

Crack in pipe A cracked may be formed due to a stress-induced dissociation on the metal surface. The crack type of damage in a pipe is displayed in Fig. 5.3.

Gouge in pipe A gouge may be caused by mechanical or forceful removal of metal from the pipe wall. The gouge type of damage in a pipe is displayed in Fig. 5.4.

Free span Free span on a submarine pipeline systems is a condition when the seabed fouling have been corroded such that the pipe systems are not anymore supported on the sea bottom as shown in Fig. 5.5.

Pipe buckling Buckling is considered as a structural instability and can create a gross deformation of the pipe cross-section as shown in Fig. 5.6.

Pipe coating Coating damage is most likely to occur when a pipe being pulled through an excavation. The coating type of damage in a pipe is displayed in Fig. 5.7.

Anode damage Anodes can be damaged when pipeline systems travel over the stinger on the rear of the lay barge. The anode deterioration on pipe is displayed in Fig. 5.8.

S. Razvarz et al., *Flow Modelling and Control in Pipeline Systems*, Studies in Systems, Decision and Control 321, https://doi.org/10.1007/978-3-030-59246-2_5

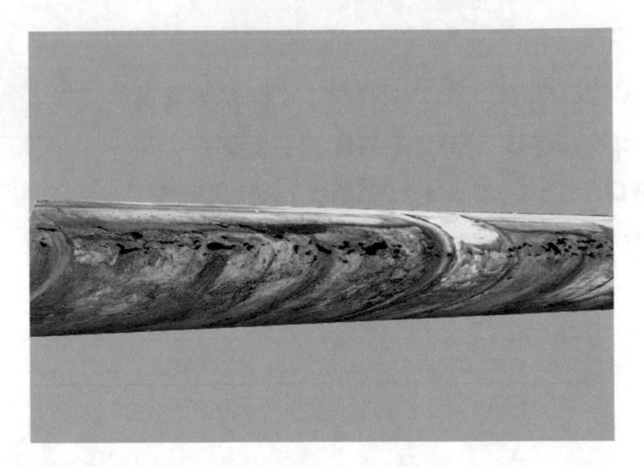

Fig. 5.1 Metal loss in pipe

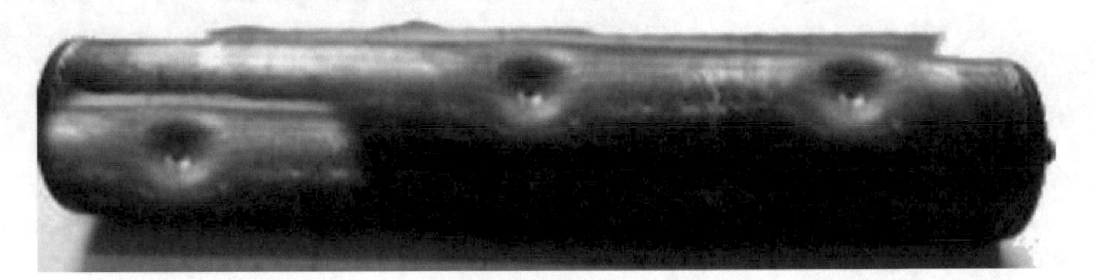

Fig. 5.2 A dent in pipe

Fig. 5.3 Crack in pipe

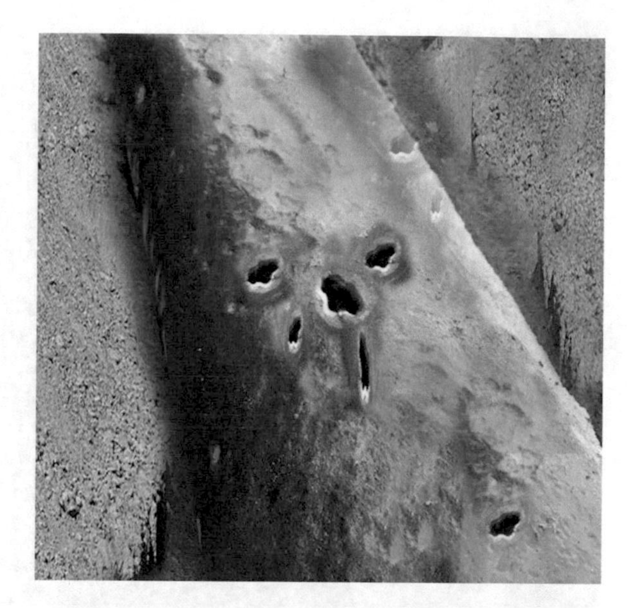

Fig. 5.4 Gouge in pipe

Fig. 5.5 Free span on a pipe

Fig. 5.6 Pipe buckling

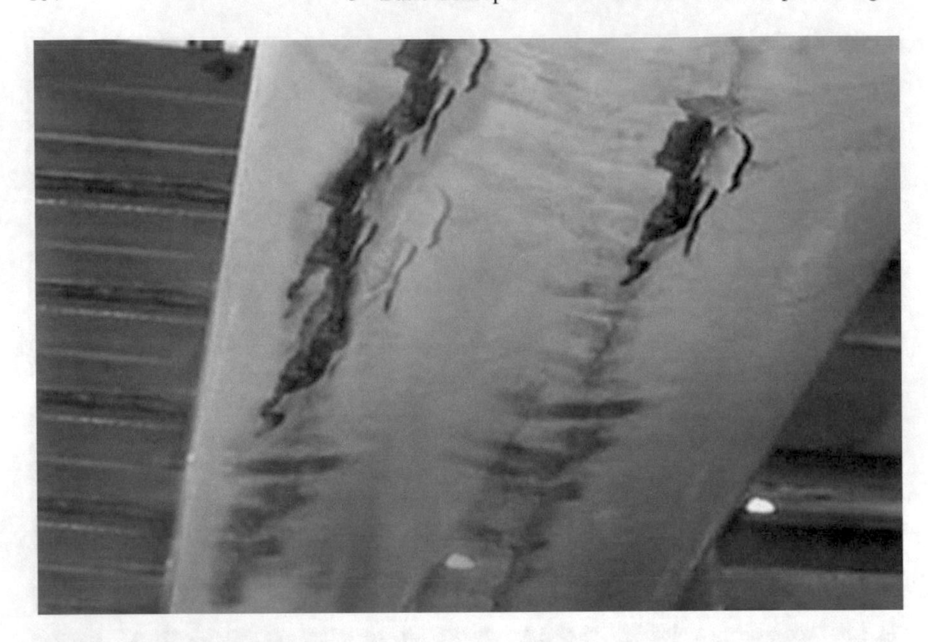

Fig. 5.7 Coating damage in pipe

Fig. 5.8 Anode deterioration on pipe

5.2 Neural Systems

Neural networks could be considered as a part of the broader field of the soft computing [1–7]. Neural networks are typically built up of neurons and synapses and alter their rates in reply from nearby neurons as well as synapses. The way neural networks work is similar to computer in input-output mapping. Neurons, as well as synapses are considered as silicon members by mimicking their treatment. A neuron receives and processes the entire incoming signals from other neurons and presents its reply by a number. Signals could travel among the synapses and have numerical features. Learning process of a neural network starts when they vary the value of their synapsis. A basic structure of a biological neuron also known as nerve cell is illustrated in Fig. 5.9. The processing steps inside an artificial neuron is illustrated in Fig. 5.10.

In this section a number of intelligent techniques used for damage detection are introduced.

Artificial neural networks Artificial neural networks have been viewed as the most widely used technique in recent two decades and have been widely applied for variety of problems in different fields [8–17]. Artificial neural networks have been emerged

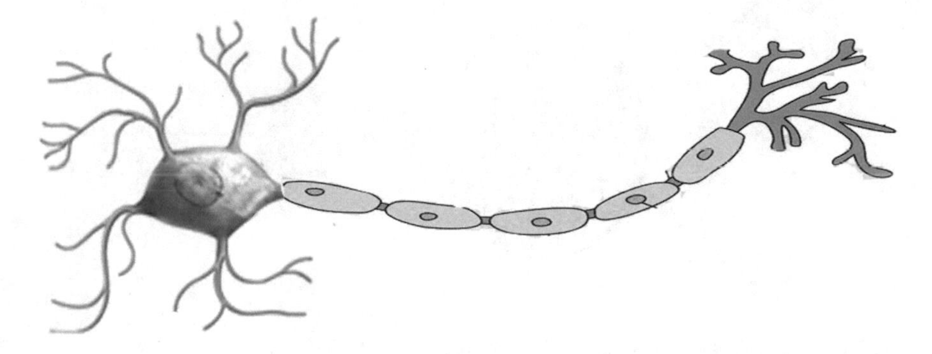

Fig. 5.9 A basic structure of a biological neuron

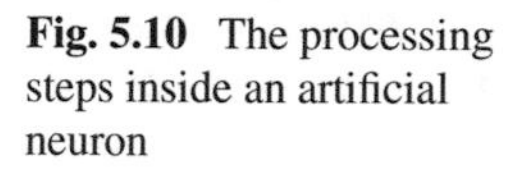

Fig. 5.10 The processing steps inside an artificial neuron

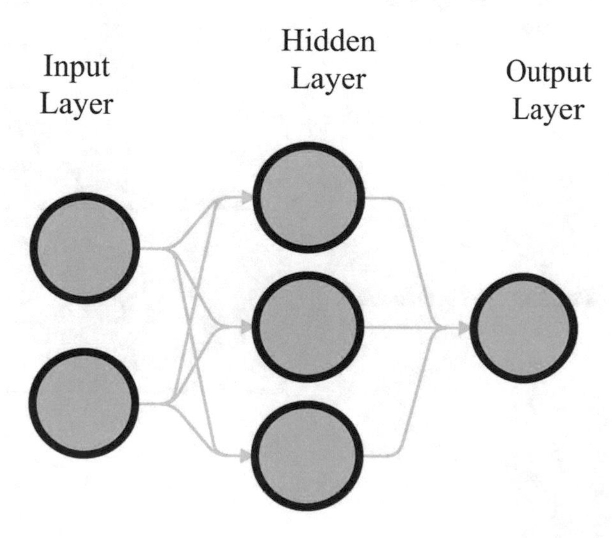

as powerful and versatile tools. They are model-free approach and are capable of learning the problems [18–22]. The artificial neuron model was initially introduced by McCulloch-Pitts in 1940 which later implemented in different feed-forward artificial neural networks like multilayer perceptrons [23]. The artificial neuron can efficiently perform a linear transformation by calculating the weighted sum using the scalar weights [24, 25]. According to Fig. 5.11, the activation function can be computed by sum of the multiplication of inputs with their corresponding weights ($k_\iota u_\iota$). In addition, the bias b can be added to the addition outcome, z. Subsequently, the new Z suits into the activation function $\psi(Z)$. Ultimately, the outcome of the activation function can be used for predicting the output neuron O (Figs. 5.12, 5.13 and 5.14).

Learning process of the artificial neural network depends on the interconnection of neurons with each other. Artificial neurons obtain either excitatory or inhibitory

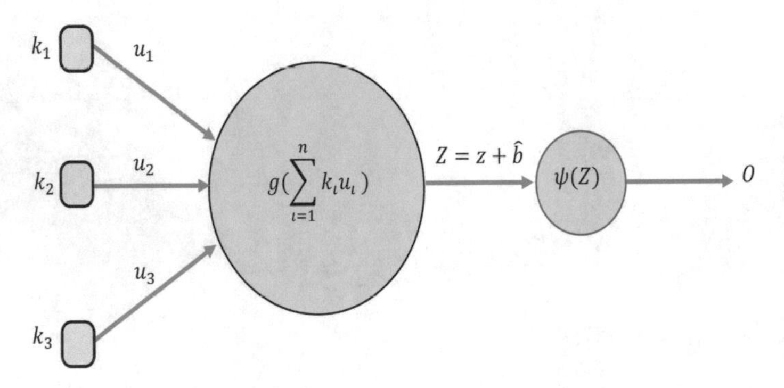

Fig. 5.11 The basic structure of an artificial neuron

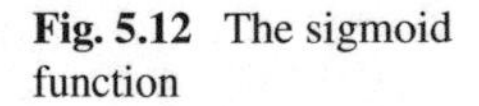

Fig. 5.12 The sigmoid function

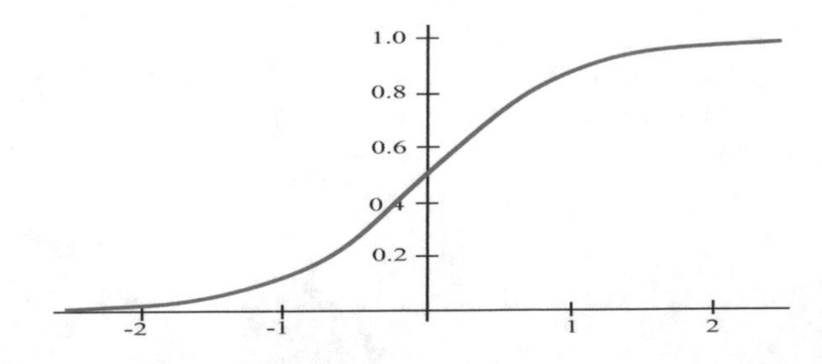

Fig. 5.13 The tanh function

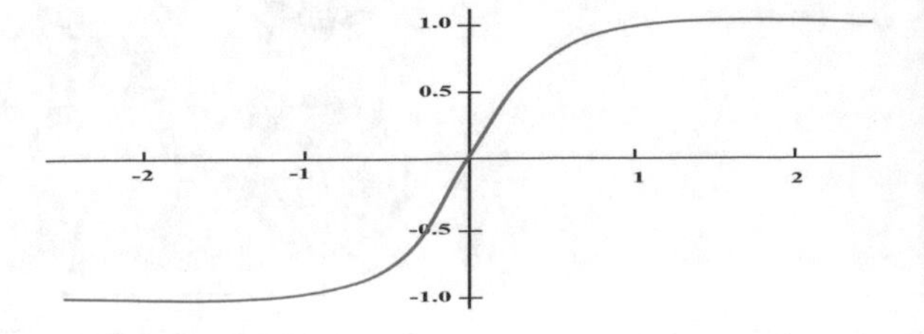

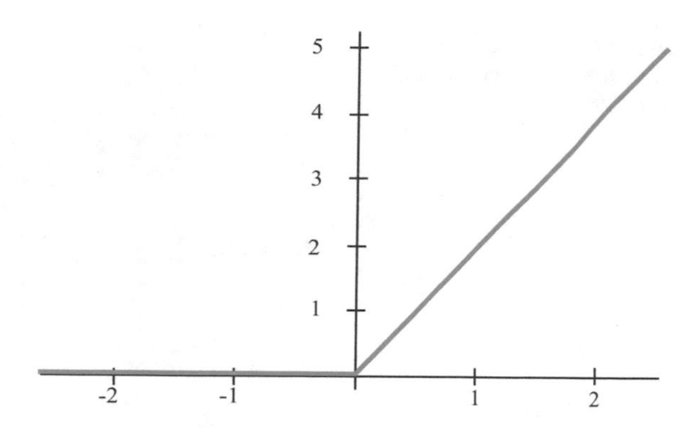

Fig. 5.14 The ReLU function

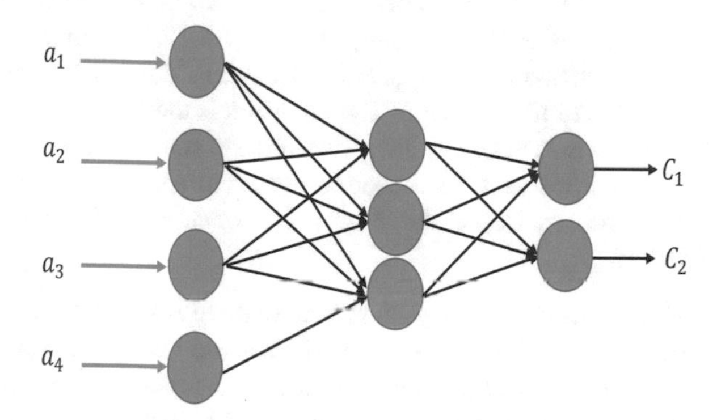

Fig. 5.15 The model structure of feedforward neural network

inputs just as biological neurons do. Excitatory inputs make the adding device of the subsequent neuron to sum whilst the inhibitory inputs make it to minus.

Feedback is a type of linkage and occurs when there is a feedback loop from the output layer to the input of a prior layer, or even to the equal layer [26, 27]. A feedforward neural network is another type of artificial neural network in which outputs cannot fed back into to the input neurons. As a result, it won't have any recording of its former output amounts, see Fig. 5.15.

The model structure of feedback neural network includes connection from output to input neurons. Each neuron generally contains one extra weight such as an input, which permits an extra grade of freedom during minimizing the training error as shown in Fig. 5.16. Feedback neural network can retain the historical record of former state, thus subsequent state besides relying on input signals is dependent on the prior states of the network.

5.3 Memory Networks

End-to-End Memory Network with Single Computational Step The end-to-end Memory Network (N2N) with single computational step termed as single hop

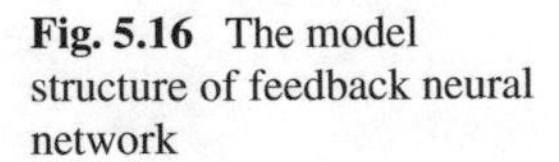
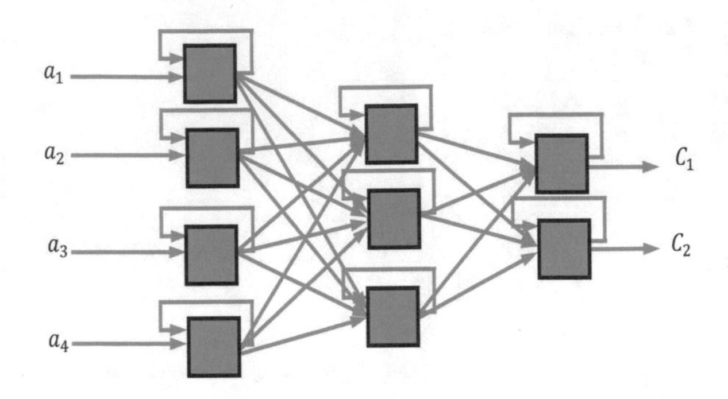

Fig. 5.16 The model structure of feedback neural network

includes two stories embedding S, G, as well as a question embedding K. The model structure of N2N is displayed in Fig. 5.17. For matching each word in the story with each word in the question, matrices dot product have been implemented and established the attention. The attention is then passed through a softmax layer and is changed into the probability distribution across the whole word from the story. These probabilities have been exerted to the story embedding G and finally the sum of that with the question embedding K has been passed through a dense and the softmax layers.

End-to-End Memory Network with Stacked Computational Steps Supporting memories and final answer prediction are two important components of N2N model [27]. Supporting memories as a first component have been made of a set of input and output memory and demonstrated by memory cells. In intricate tasks that they may need multiple supporting memories, a new structure of N2N can be made which contains more than one set of input-output memories by stacking a number of memory

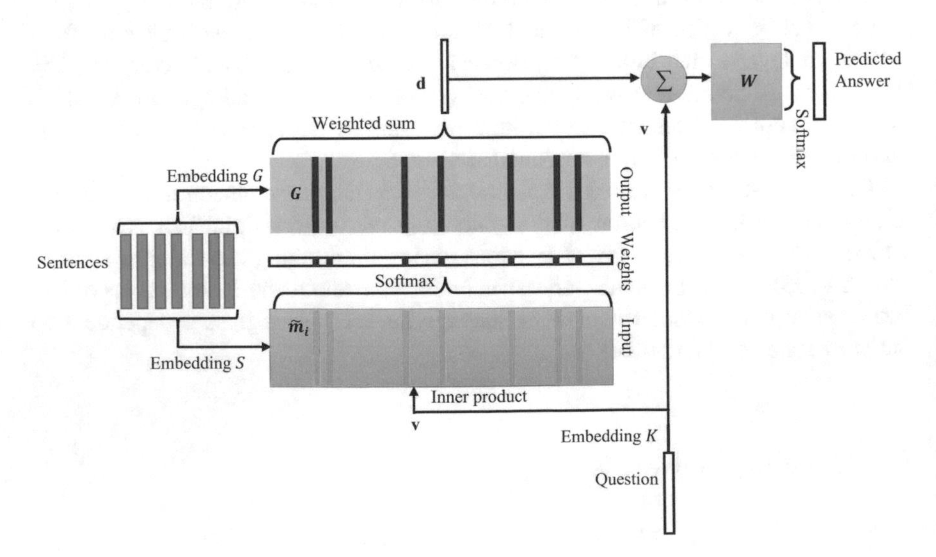

Fig. 5.17 End-to-end memory network with single computational step

layers. In N2N with stacked computational steps each memory layer is named as hop besides the input of the (h + 1)th hop is the output of the hth hop:

$$v^{h+1} = d^h + v^h \tag{5.1}$$

In this model each layer contains its own embedding matrices S^h, G^h, implemented for embedding the inputs.

The predicted answer to the question can be calculated as follows,

$$z = \text{softmax}\left(W\left(d^h + v^h\right)\right) \tag{5.2}$$

in which W is a parameter matrix for the N2N with stacked computational steps in order to learn, and also h is the total number of hops.

Figure 5.18 demonstrates the model structure of N2N with three hop operations. In this model, the hard max operations defined in each layer have been substituted with a continuous weighting from the softmax. The N2N with stacked computational steps takes a discrete set of inputs $k_1, \ldots, k_n$ stored in the memory and a question b, and generates a reply y. All k can be written to the memory up to a fixed buffer size, besides a continuous demonstration for k and b can be obtained. The continuous demonstration have been processed with multiple hops for producing y which permits backpropagation of the error signal through multiple memory accesses back into the input during the training process.

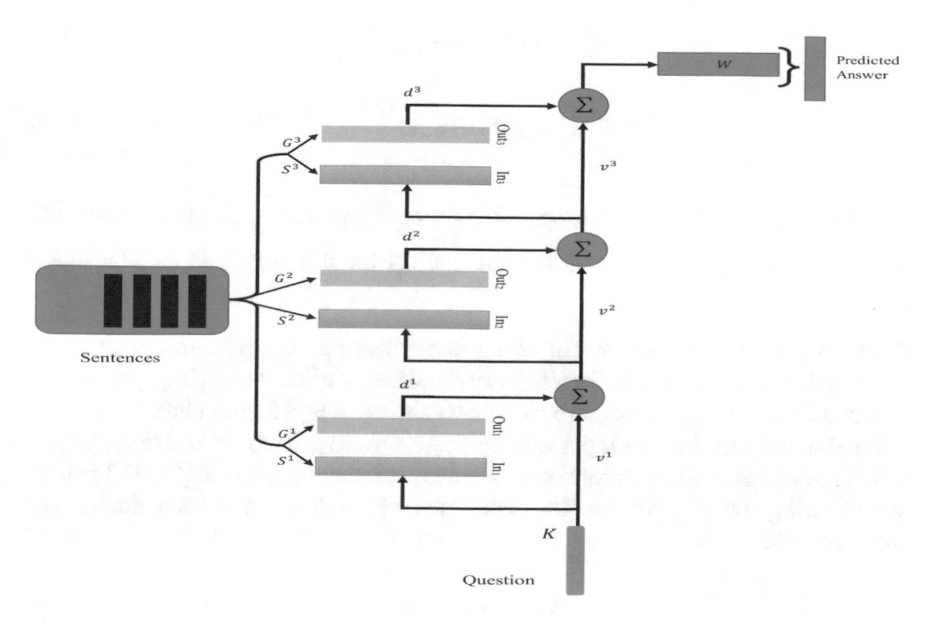

Fig. 5.18 The model structure of a three layer end-to-end memory network

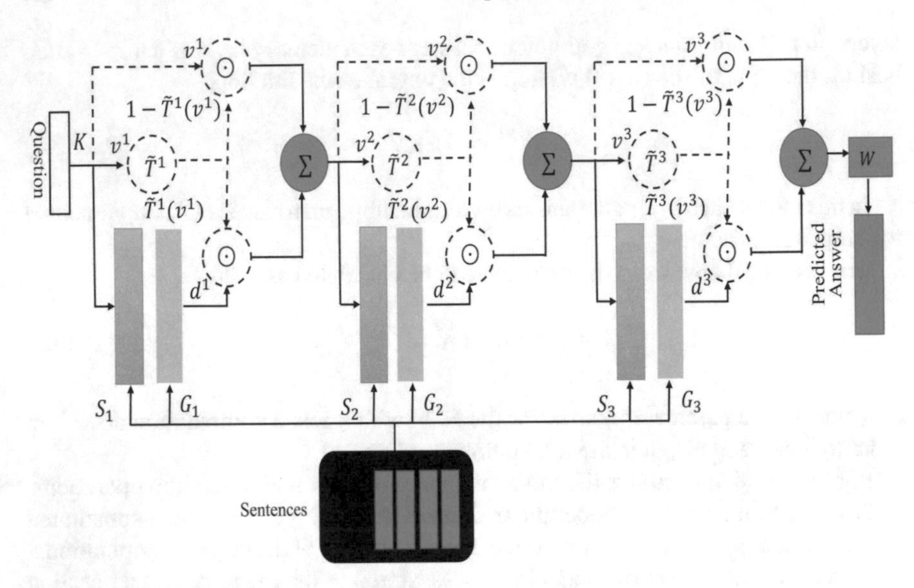

Fig. 5.19 The model structure of a gated end-to-end memory network

Gated End-to-End Memory Network As described in Fig. 5.19, the gated end-to-end memory network (GN2N) can dynamically initiate memory read on the controller state v^h at every hop. In GN2N model, (5.1) can be described as follows [28],

$$\tilde{T}^h\left(v^h\right) = \sigma\left(W^h_{\tilde{T}} v^h + a^h_{\tilde{T}}\right) \tag{5.3}$$

$$v^{h+1} = d^h \odot \tilde{T}^h\left(v^h\right) + v^h \odot \left(1 - \tilde{T}^h\left(v^h\right)\right) \tag{5.4}$$

such that $W^h_{\tilde{T}}$ and a^h are the hop-specific parameter matrix and bias term for the hth hop respectively. $\tilde{T}^h(k)$ is the transform gate for the hth hop. $\odot$ is the Hadamard product.

Fuzzy Associative Memory The fuzzy associative memory system can be a two layer feedforward, hetroassociative, fuzzy classifier, or a fuzzy mapping system [29]. The structure of fuzzy associative memory is shown in Fig. 5.20. In Fig. 5.20, C is a s-dimensional fuzzy vector input variable that activates p fuzzy rules also is mapped into a m-dimensional fuzzy vector D. Numerous fuzzy vector pair (C_r, D_r) can be stored during learning process. The demonstrations of these variables in the form of fuzzy sets are as below

$$\begin{gathered} C_r = \left(c_1^r, \ldots, c_s^r\right) \\ D_r = \left(d_1^r, \ldots, d_m^r\right) \\ G_r = C_0^T D = \min\left[\mu_C\left(c_i^r\right), \mu_D\left(d_j^r\right)\right] \end{gathered} \tag{5.5}$$

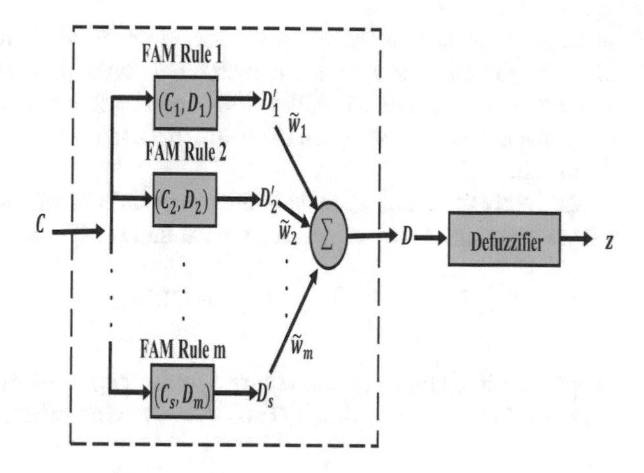

Fig. 5.20 Fuzzy associative memory model

where $\mu C(c_i^r)e$ is the membership of ith element in fuzzy vector C, $\mu_D\left(d_j^r\right)$ is the membership of jth element in fuzzy vector D, o is a composition operator, also Gr is a s × m matrix. In fuzzy associative memory model, (C_r, D_r) presents the rth rule, also signifies "If W is C_r Then Z is D_r." The fuzzy rule of FAM can be in the form of $(C_r, D_r; H_r)$ that signifies "If W is C_r and Z is D_r Then V is H_r."

End-to-End Memory Networks with Unified Weight Tying Two kinds of weight tying approaches namely adjacent and layer-wise can be defined for N2N [27]. Layer-wise technique can portion the input and output embedding matrices across multiple hops (i.e., $S^1 = S^2 = \ldots = S^h$ and $G^1 = G^2 = \ldots = G^h$). Adjacent technique can portion the output embedding for a supplied layer with the corresponding input embedding (i.e., $S^{h+1} = G^h$). Besides, the matrix W for estimating the answer, and the matrix K for question embedding can be defined as $W^{\tilde{T}} = G^h$ and $K = S^1$. In [28], a dynamic mechanism is designed which permits the model to choose the proper kind of weight tying on the basis of the input. In [28], the researchers proposed a dynamic model that could choose the proper kind of weight tying through the use of the input. The embedding matrices in UN2N have been generated dynamically for every instance and makes this model more efficient when is compared with N2N and GN2N since the same embedding matrices have been used for each input in those models. A gating vector x can be applied for generating the embedding matrices S^h, G^h, K, and W. These embedding matrices have been impacted by the information transported utilizing x in connection with the input question υ_0 and also the context sentences in the story $\tilde{m}_t$, thus,

$$S^{h+1} = S^h \odot x + G^h \odot (1 - x) \tag{5.6}$$

$$G^{h+1} = G^h \odot x + G^{h+1} \odot (1 - x) \tag{5.7}$$

in which $\odot$ is considered as the column element-wise multiplication operation, and also G^{h+1} is taken to be the unconstrained embedding matrix. By choosing a large value of x in (5.6) and (5.7), the UN2N can be stated as layer-wise approach and by choosing a small value of x in (5.6) and (5.7), the UN2N can be stated as the adjacent approach.

In UN2N model as a first step the story has been encoded by reading the memory one step at a time with a gated recurrent unit (GRU) which is defined as follows,

$$k_{t+1} = GRU(\tilde{m}_t, k_t) \tag{5.8}$$

in which t is taken to be the recurrent time step, also $\tilde{m}_t$ is taken to be the context sentence in the story at time t. Thus the following relation has been given,

$$x = \sigma\left(W_x \begin{bmatrix} v^0 \\ k_{\tilde{T}} \end{bmatrix} + a_x\right) \tag{5.9}$$

such that $k_{\tilde{T}}$ is taken to be the last hidden state of the GRU and demonstrates the story, W_x is taken to be the weight matrix, a_x is taken to be the bias, σ is taken to be the sigmoid function also, $\begin{bmatrix} v^0 \\ k_{\tilde{T}} \end{bmatrix}$ is taken to be the concatenation of v^0 and $k_{\tilde{T}}$. A linear mapping $J \in R^{d \times d}$ has been added for updating the connection among memory hops as below,

$$v^{h+1} = d^h + (J \odot (1 - x))v^h \tag{5.10}$$

In order to take a useful action while repairing structural damage it is essential to understand the kind of damage. It has been reported that the great number of infrastructures in the world have been aged and also the the number of structures that should be investigated is growing fast within the next few years. End-to-end memory networks as an efficient technique can be used for damage detection. In [30] an end-to-end approach for object detection which depends on deep learning has been proposed to the road surface damage identification.

Recurrent Neural Networks Another efficient technique named as recurrent neural network has been proposed for dealing with sequences of data [31]. This model can save the old information of the prior sequences and transits it to the subsequent sequences [32–34]. Figure 5.21 displays a simple recurrent neural network in which at is input, y_t is output and j is a loop between them.

The information from a recurrent cell can be saved using j Loop and can be crossed to the subsequent cell in the subsequent recurrent network.

The models for leak detection can be divided into two groups: internal and external models. Internally based models for leak detection utilize flow, pressure as well as fluid temperature in internal monitoring for pipeline. Externally based models for

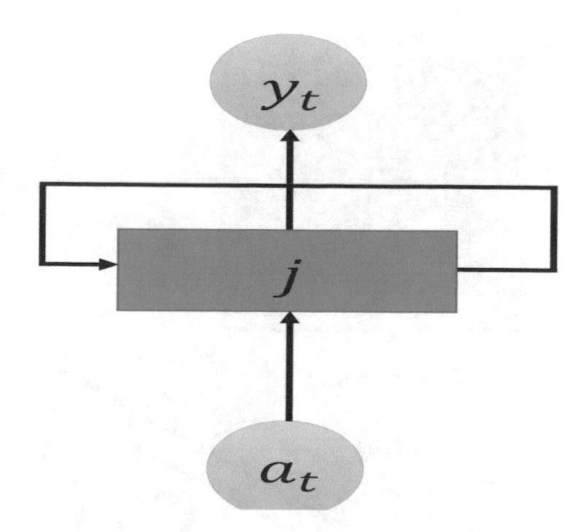

Fig. 5.21 The model structure of a recurrent neural network

leak detection utilize local dedicated sensors in external monitoring for pipeline. Recurrent neural networks have been efficiently utilized for nonlinear system identification for fuel oil leak detection [35]. The data for training the recurrent neural networks can be obtained from inlet and outlet flow parameters of the pipeline.

Convolutional Neural Networks The structure of convolutional neural network is based on convolutional and pooling layers [36–39]. In the convolutional neural networks two learning procedures named the feature extraction and feature classification have been combined in such a manner that a sole learning frame has been produced. Convolutional neural networks can easily adapt to various input sizes and also have the ability to learn how to optimize the features while training stage straightly from the raw input.

The model structure of a convolutional neural network made of two convolution and one fully-connected layers has been demonstrated in Fig. 5.22. In Fig. 5.22 a 2424-pixel grayscaling image has been categorized into two classes.

The acoustic signals derived by an audition tool on pipeline systems can be utilized to detect the leakage in buried water pipe networks. Basically, a listening machine can be used for gathering the sonic signal along the path of pipelines and validates the sonic signal if a leak is in the pipelines or not. Since in this technique the leak signals are constantly corrupting with non-leak sonic sources, therefore it will be difficult to identify leakage and also it may need more time to locate the leakage. However, convolutional neural networks can be efficiently utilized for leak identification in pipelines [40] as the convolutional neural network has high classification accuracy.

Fuzzy Systems Fuzzy system is an extremely appropriate tool for estimated reasoning, especially for the system with a mathematical model which is quite difficult to achieve [41–52]. systems engineers face particular challenges with inadequate and unsure information [53]. Fuzzy logic theory has been efficiently used for resolving these kinds of problems. Moreover, fuzzy systems can be utilized for finding multiple leaks in a pipeline [54, 55].

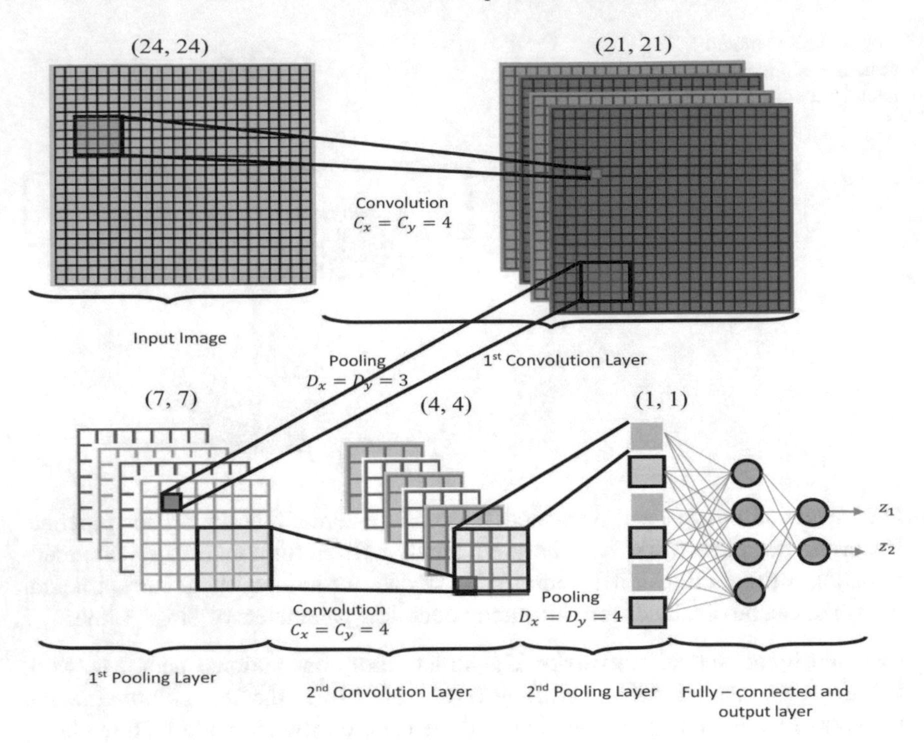

Fig. 5.22 The model structure of a convolutional neural network

5.4 Applications

In this section, some real-life examples have been utilized to demonstrate the application of neural networks and deep learning models in detecting damaged pipelines.

5.4.1 *Example 0.1*

Artificial neural network approach can be utilized for finding multiple leaks in a pipeline [56]. The model of the fluid flow through a pipe as shown in Fig. 5.23 can be stated as following,

$$\frac{\partial H}{\partial t} + qD\frac{\partial K}{\partial u} + \varphi|H|H = 0,\ z^2\frac{\partial H}{\partial u} + qD\frac{\partial K}{\partial t} = 0 \tag{5.11}$$

in which K is considered as the pressure head, H is considered as the flow, u is considered as the length coordinate, t is considered as the time coordinate, q is taken

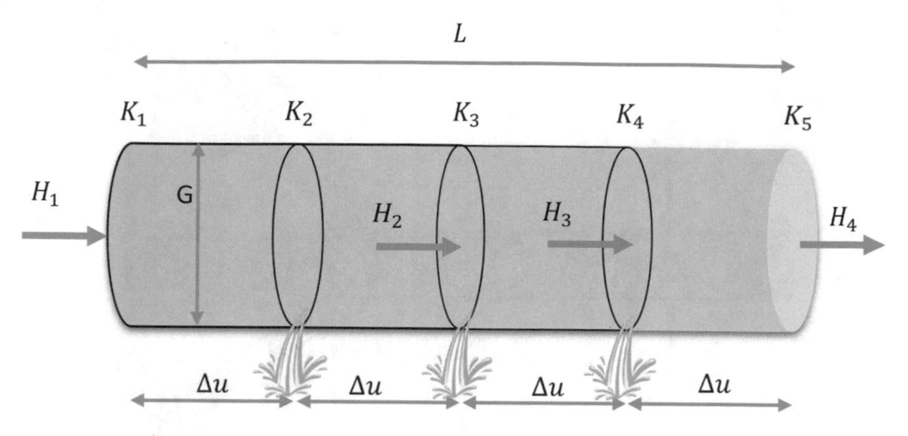

Fig. 5.23 The model of the fluid flow through a pipe

to be the acceleration of the gravity, D is taken to be the cross-section area, z is considered as the speed of sound, and $\varphi = a/2GD$. G is taken to be the pipeline diameter, and a is taken to be the Darcy-Weissbach friction coefficient.

A precise leak detector system consists of an artificial neural network can be used in finding the leak location in the pipeline. The precision of the leak detector system is in direct relation with the performance of the artificial neural network. The inlet/outlet flow can be measured using artificial neural network. The detector system distinguishes the possible pipeline operating states that can be utilized to detect and pinpoint the location of leaks.

5.4.2 *Example 0.2*

In Fig. 5.24 the water in the pipe a_1 has been partitioned into three pipes a_2, a_3 and a_4. The pipes have uncertain area which are $B_1 = Q(0.3, 0.5, 0.7, 0.9)$, $B_2 = Q(0.07, 0.09, 0.3, 0.5)$, $B_3 = Q(0.05, 0.09, 0.3, 0.4)$. The water velocities in the pipes a_1, a_2, a_1 and a are $\vartheta_1 = s^5$, $\vartheta_2 = \frac{e^s}{3}$, $\vartheta_3 = s$ respectively and can be controlled by the valves parameter s. The control object is to allow liquid to flow though pipes

$$P = Q(11.653219,\ 13.986834,\ 15.142741,\ 17.325734) \tag{5.12}$$

The aim is to find the valve control parameter s. The mass balance formula can be defined as follows

$$B_1\vartheta_1 = B_2\vartheta_2 \oplus B_3\vartheta_3 \oplus P \tag{5.13}$$

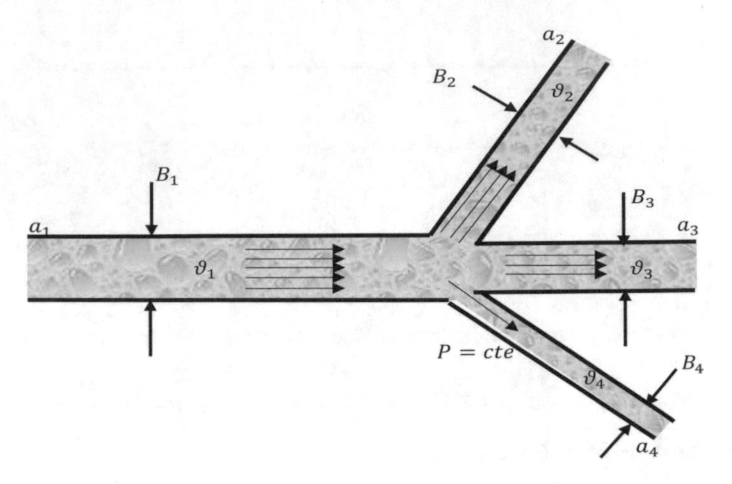

Fig. 5.24 Water-channel system

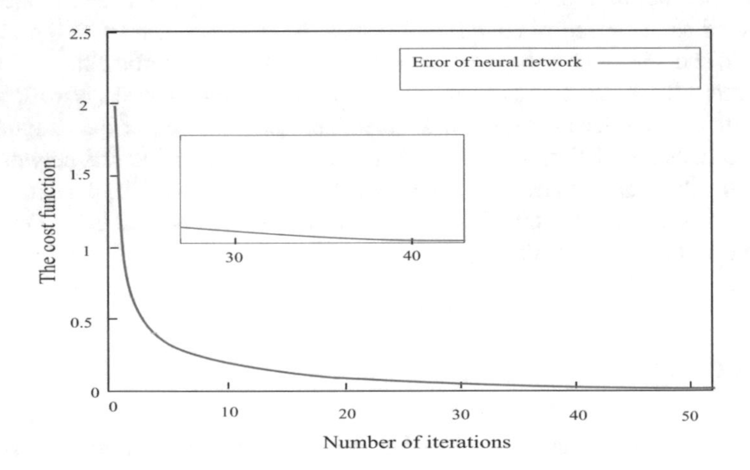

Fig. 5.25 The graph of the error

The exact solution is $s = 4$ and the maximum learning rate related to the artificial neural network approach is taken to be $\eta = 0.002$. The graph of the error is given in Fig. 5.25.

References

1. Jafari, R., Razvarz, S., Gegov, A., Yu, W.: Fuzzy control of uncertain nonlinear systems with numerical techniques: a survey. In: UK Workshop on Computational Intelligence, pp. 3–14. Springer (2019)

2. Jafari, R.: A new iterative approach based on artificial intelligence for solving dual fuzzy polynomials (2016)
3. Jafarian, A., Jafari, R.: Simulation and evaluation of fuzzy polynomials by feed-back neural networks (2012)
4. Jafarian, A., Jafari, R.: An iterative method for solving fuzzy polynomials by fuzzy neural networks (2012)
5. Jafari, R., Yu, W.: Artificial neural network approach for solving strongly degenerate parabolic and burgers-fisher equations. In: 2015 12th International Conference on Electrical Engineering, Computing Science and Automatic Control (CCE), pp. 1–6. IEEE (2015)
6. Jafari, R., Yu, W.: Uncertainty nonlinear systems control with fuzzy equations. In: 2015 IEEE International Conference on Systems, Man, and Cybernetics, pp. 2885–2890. IEEE (2015)
7. Jafari, R., Yu, W.: Uncertainty nonlinear systems modeling with fuzzy equations. In: 2015 IEEE International Conference on Information Reuse and Integration, pp. 182–188. IEEE (2015)
8. Jafari, R., Razvarz, S., Gegov, A., Paul, S.: Modeling and control of uncertain nonlinear systems. In: 2018 International Conference on Intelligent Systems (IS), pp. 168–173. IEEE (2018)
9. Jafari, R., Razvarz, S., Yu, W., Gegov, A., Goodwin, M., Adda, M.: Genetic algorithm modeling for photocatalytic elimination of impurity in wastewater. In: Proceedings of SAI Intelligent Systems Conference, pp. 228–236. Springer (2019)
10. Jafari, R., Yu, W.: Uncertain nonlinear system control with fuzzy differential equations and Z-numbers. In: 2017 IEEE International Conference on Industrial Technology (ICIT), pp. 890–895. IEEE (2017)
11. Jafari, R., Yu, W., Li, X.: Solving fuzzy differential equation with Bernstein neural networks. In: 2016 IEEE International Conference on Systems, Man, and Cybernetics (SMC), pp. 001245–001250. IEEE (2016)
12. Jafarian, A., Jafari, R.: Approximate solutions of dual fuzzy polynomials by feed-back neural networks. J. Soft Comput. Appl. **2012**, 1–5 (2012)
13. Jafarian, A., Jafari, R.: New method for solving fuzzy polynomials. Adv. Fuzzy Math. **8**(1), 25–33 (2013)
14. Jafarian, A., Measoomy, N.S., Jafari, R.: Solving fuzzy equations using neural nets with a new learning algorithm (2012)
15. Razvarz, S., Jafari, R.: ICA and ANN modeling for photocatalytic removal of pollution in wastewater. Math. Comput. Appl. **22**(3), 38 (2017)
16. Razvarz, S., Jafari, R.: Intelligent techniques for photocatalytic removal of pollution in wastewater. J. Electr. Eng. **5**(1), 321–328 (2017)
17. Razvarz, S., Jafari, R., Yu, W., Golmankhaneh, A.K.: PSO and NN modeling for photocatalytic removal of pollution in wastewater. In: 2017 14th International Conference on Electrical Engineering, Computing Science and Automatic Control (CCE), pp. 1–6. IEEE (2017)
18. Jafari, R., Razvarz, S., Gegov, A., Paul, S.: Fuzzy modeling for uncertain nonlinear systems using fuzzy equations and Z-numbers. In: UK Workshop on Computational Intelligence, pp. 96–107. Springer (2018)
19. Jafari, R., Razvarz, S., Gegov, A.: A new computational method for solving fully fuzzy nonlinear systems. In: International Conference on Computational Collective Intelligence, pp. 503–512. Springer (2018)
20. Razvarz, S., Jafari, R., Gegov, A.: Solving partial differential equations with Bernstein neural networks. In: UK Workshop on Computational Intelligence, pp. 57–70. Springer (2018)
21. Razvarz, S., Jafari, R., Gegov, A., Paul, S.: Neural network approach to solving fully fuzzy nonlinear systems. In: Fuzzy Modeling and Control: Methods, Applications and Research, pp. 46–68. Nova Science Publishers, Inc. (2018)
22. Razvarz, S., Jafari, R., Granmo, O.-C., Gegov, A.: Solution of dual fuzzy equations using a new iterative method. In: Asian Conference on Intelligent Information and Database Systems, pp. 245–255. Springer (2018)
23. McCulloch, W.S., Pitts, W.: A logical calculus of the ideas immanent in nervous activity. Bull. Math. Biophys. **5**(4), 115–133 (1943)

24. Jafari, R., Razvarz, S., Gegov, A.: Fuzzy differential equations for modeling and control of fuzzy systems. In: International Conference on Theory and Applications of Fuzzy Systems and Soft Computing, pp. 732–740. Springer (2018)
25. Jafari, R., Razvarz, S., Gegov, A.: Neural network approach to solving fuzzy nonlinear equations using Z-numbers. IEEE Trans. Fuzzy Syst. (2019)
26. Jafarian, A., Jafari, R., Golmankhaneh, A.K., Baleanu, D.: Solving fully fuzzy polynomials using feed-back neural networks. Int. J. Comput. Math. **92**(4), 742–755 (2015)
27. Sukhbaatar, S., Weston, J., Fergus, R.: End-to-end memory networks. In: Advances in Neural Information Processing Systems, pp. 2440–2448 (2015)
28. Liu, F., Cohn, T., Baldwin, T.: Improving end-to-end memory networks with unified weight tying. In: Proceedings of the Australasian Language Technology Association Workshop 2017, pp. 16–24 (2017)
29. Wang, W., Wu, G.F., Huang, B.S., Zhuang, K.Y., Zhou, P.L., Jiang, C.X., Li, D.S., Zhou, Y.H.: The FAM (fuzzy associative memory) neural network model and its application in earthquake prediction. Acta Seismol. Sin. **10**(3), 321–328 (1997)
30. Maeda, H., Sekimoto, Y., Seto, T., Kashiyama, T., Omata, H.: Road damage detection using deep neural networks with images captured through a smartphone (2018). arXiv preprint arXiv: 180109454
31. Elman, J.L.: Distributed representations, simple recurrent networks, and grammatical structure. Mach. Learn. **7**(2–3), 195–225 (1991)
32. Pearlmutter, B.A.: Learning state space trajectories in recurrent neural networks. Neural Comput. **1**(2), 263–269 (1989)
33. Robinson, T., Hochberg, M., Renals, S.: The use of recurrent neural networks in continuous speech recognition. In: Automatic Speech and Speaker Recognition, pp. 233–258. Springer (1996)
34. Perez, J., Liu, F.: Gated end-to-end memory networks (2016). arXiv preprint arXiv:161004211
35. Mohammadi, M., Nikbakht, A., Bavalishoar, A.: Fuel oil leak detection in power plant with recurrent neural network and execute in programmable logic controller. In: 2015 2nd International Conference on Knowledge-Based Engineering and Innovation (KBEI), pp. 927–932. IEEE (2015)
36. LeCun, Y., Bottou, L., Bengio, Y., Haffner, P.: Gradient-based learning applied to document recognition. Proc. IEEE **86**(11), 2278–2324 (1998)
37. Jarrett, K., Kavukcuoglu, K., Ranzato, M.A., LeCun, Y.: What is the best multi-stage architecture for object recognition? In: 2009 IEEE 12th International Conference on Computer Vision, pp. 2146–2153. IEEE (2009)
38. Lee, H., Grosse, R., Ranganath, R., Ng, A.Y.: Convolutional deep belief networks for scalable unsupervised learning of hierarchical representations. In: Proceedings of the 26th Annual International Conference on Machine Learning, pp. 609–616 (2009)
39. Turaga, S.C., Murray, J.F., Jain, V., Roth, F., Helmstaedter, M., Briggman, K., Denk, W., Seung, H.S.: Convolutional networks can learn to generate affinity graphs for image segmentation. Neural Comput. **22**(2), 511–538 (2010)
40. Chuang, W.-Y., Tsai, Y.-L., Wang, L.-H.: Leak detection in water distribution pipes based on CNN with mel frequency cepstral coefficients. In: Proceedings of the 2019 3rd International Conference on Innovation in Artificial Intelligence, pp. 83–86 (2019)
41. Jafari, R., Razvarz, S.: Solution of fuzzy differential equations using fuzzy Sumudu transforms. Math. Comput. Appl. **23**(1), 5 (2018)
42. Jafari, R., Yu, W.: Fuzzy modeling for uncertainty nonlinear systems with fuzzy equations. Math. Probl. Eng. (2017)
43. Jiang, W., Xie, C., Luo, Y., Tang, Y.: Ranking Z-numbers with an improved ranking method for generalized fuzzy numbers. J. Intell. Fuzzy Syst. **32**(3), 1931–1943 (2017)
44. Negoiţă, C.V., Ralescu, D.A.: Applications of Fuzzy Sets to Systems Analysis. Springer (1975)
45. Tatchum, M., Gegov, A., Jafari, R., Razvarz, S.: Parallel distributed compensation for voltage controlled active magnetic bearing system using integral fuzzy model. In: 2018 International Conference on Intelligent Systems (IS), pp. 190–198. IEEE (2018)

46. Wakami, N., Araki, S., Nomura, H.: Recent applications of fuzzy logic to home appliances. In: Proceedings of IECON'93–19th Annual Conference of IEEE Industrial Electronics, pp. 155–160. IEEE (1993)
47. Yaakob, A.M., Gegov, A.: Fuzzy rule based approach with z-numbers for selection of alternatives using TOPSIS. In: 2015 IEEE International Conference on Fuzzy Systems (FUZZ-IEEE), pp. 1–8. IEEE (2015)
48. Zadeh, L.A.: Probability measures of fuzzy events. J. Math. Anal. Appl. **23**(2), 421–427 (1968)
49. Zadeh, L.A.: Calculus of fuzzy restrictions. In: Fuzzy Sets and Their Applications to Cognitive and Decision Processes, pp. 1–39. Elsevier (1975)
50. Zamri, N., Ahmad, F., Rose, A.N.M., Makhtar, M.: A fuzzy TOPSIS with Z-numbers approach for evaluation on accident at the construction site. In: International Conference on Soft Computing and Data Mining, pp. 41–50. Springer (2016)
51. Jafari, R., Razvarz, S., Gegov, A.: A novel technique for solving fully fuzzy nonlinear systems based on neural networks. Vietnam J. Comput. Sci. **7**(1), 93–107 (2020)
52. Abiyev, R.H., Uyar, K., Ilhan, U., Imanov, E., Abiyeva, E.: Estimation of food security risk level using Z-number-based fuzzy system. J. Food Qual. (2018)
53. Nuriyev, A.: Application of Z-numbers based approach to project risks assessment. Eur. J. Interdiscip. Stud. **5**(2), 67–73 (2019)
54. Da Silva, H.V., Morooka, C.K., Guilherme, I.R., da Fonseca, T.C., Mendes, J.R.: Leak detection in petroleum pipelines using a fuzzy system. J. Petrol. Sci. Eng. **49**(3–4), 223–238 (2005)
55. Feng, J., Zhang, H., Liu, D.: Applications of fuzzy decision-making in pipeline leak localization. In: 2004 IEEE International Conference on Fuzzy Systems (IEEE Cat. No. 04CH37542), pp. 599–603. IEEE (2004)
56. Barradas, I., Garza, L.E., Morales-Menendez, R., Vargas-Martínez, A.: Leaks detection in a pipeline using artificial neural networks. In: Iberoamerican Congress on Pattern Recognition, pp. 637–644. Springer (2009)

Chapter 6
Leakage Modelling for Pipeline

6.1 Introduction

The pipeline is a cost-effective and reliable way of transporting huge quantities of crude oil, oil products, and gas over long distances. It is one of the main objectives of the oil and gas companies to minimize the risk of leakage, spill, and theft that happen. Pipeline leak detection systems can be utilised to determine where a leak has occurred in pipeline networks and notify the operator in case of leakage incidents. It is essential that pipeline operators have a leak detection system so that they can detect leaks quickly and have rapid responses to stop oil discharge and suitable pipeline maintenance. There are several different available approaches potentially suitable for leak detection and location in pipelines based on the nature of the fluid being transported as well as the size of leaks. These techniques can start from basic designs such as mass balance approaches and extend to more complex systems. Some of the available literature on pipeline leak detection and localisation using different approaches are given in [1–13]. Artificial intelligence has become the most effective approach which attracts many investigators to deeply research [14–36]. It has been successfully used for leak detection. Other methods of leak detection include hydrostatic testing, infrared, and laser technology.

The type of damage visible on the failed pipes consisted of corrosion, dents, gouges, weld defects, metal loss, etc. These flaws could break the pipeline and result in harm to human health and the environment [37–43]. Therefore, damage detection is a necessary component of pipeline risk management. Health monitoring and accurate fault location characterize high efficiency and quality production systems. The primary purpose of the automatic pipeline monitoring system is to diagnose operational failures as fast as possible to avoid safety and environmental issues [44–46]. There are a number of methods for the straight identification of leaks in the pipeline network, including hardware-based methods and computational pipeline monitoring methods. Hardware-based techniques use hardware sensors to directly detect the occurrence of a leak that are installed along the pipelines and can mitigate risks

S. Razvarz et al., *Flow Modelling and Control in Pipeline Systems*, Studies in Systems, Decision and Control 321, https://doi.org/10.1007/978-3-030-59246-2_6

of a small leak. Although hardware-based approaches have high precision, these techniques are not economical.

Pipes are usually subjected to unwanted damages. Leakage occurs in all distribution networks and can be considered as a factor of water losses in water distribution systems. Clean and accessible water is critical for human well-being. Insufficient water supply to satisfy basic human needs has been considered as one of the top critical issues. In [47] it has been demonstrated that in the cities of the United States the average water loss in systems is 15 up to 25%. It was reported by the Canadian Water Research Institute that 20% the water provided to urban drinking water distribution systems is lost due to leakage and other unexpected factors [48]. A field study on leakage assessment was undertaken in ten areas of Riyadh, Saudi Arabia in which the average leakage was found to be about 30% [49]. Water loss due to any leaks represent an important portion of the water supply network, therefore, reducing water loss from water supply systems is critical to efficient resource utilization.

As pipeline systems are an essential mode of transporting fluid, therefore detection of leaks may be the most important challenge of pipeline networks. Several pipeline leak detection techniques have been developed during the last years which are mainly model-based techniques [50–54]. Nevertheless, computational methods for pipe networks observation rely on mathematical models of the pipeline [55]. These techniques cannot be directly applied. In this sense, they are associated with different variables such as pressure, flow rate, and temperature, between others.

Besides the detection of a leakage, the major part of the leak detection technology is to locate the accurate position of leakage. Hence, powerful algorithms can be implemented to accurately locate a leak in the pipeline [56–62]. In this chapter, we have proposed a model-based approach to fault detection in pipeline systems.

6.2 Leak Modeling

A mathematical expression for a leak can be deduced from Bernoulli's equation, which relates the pressure difference between the pipeline inside and outside.

Bernoulli's equation applies under the following considerations [63, 64].

- Non viscous flow
- Permanent or continuous flow
- Along the Pipe line
- Constant density.

Let us consider a hole in a pipe as shown in Fig. 6.1 Bernoulli's equation between points 1 and 2 is given by

$$\frac{\wp_1}{\gamma} + \frac{v_1^2}{2} + Z_1 = \frac{\wp_2}{\gamma} + \frac{v_2^2}{2} + Z_2 \tag{6.1}$$

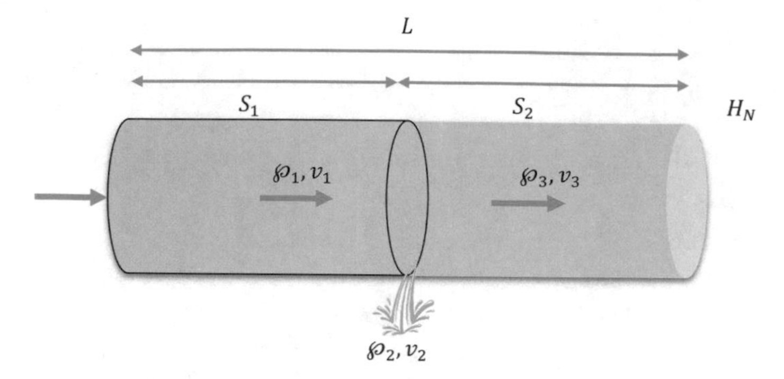

Fig. 6.1 Scheme of a hole in a pipe

where P is the pressure of the fluid load, v is the speed of the fluid, and Z is the elevation.

By taking into account that

$$Z_1 - Z_2 = 0,\ \wp_2 = \wp_{atm} = 0 \text{ and } \mathcal{H} = \frac{\wp_1}{\gamma} + \frac{v_1}{2g} \tag{6.2}$$

reduces to

$$v_2 = \sqrt{2g\mathcal{H}} \tag{6.3}$$

This represents the speed of the fluid outside the hole. Since the flow at such a point is given by the product of the output velocity and the hole area, we get

$$\mathbb{Q} = C_d \mathrm{A}_{leak}\sqrt{2g}\sqrt{\mathcal{H}(z_{leak}, t)} \tag{6.4}$$

where $\mathbb{Q}$ is the theoretical flow rate through the hole C_d is the discharge coefficient, also A_{leak} is the leak cross-section area.

Finally, the expression for the flow of a leak in a pipe at a point z_{leak} is given by

$$\mathbb{Q}_{leak} = \lambda\sqrt{\mathcal{H}(z_{leak}, t)} \tag{6.5}$$

where $\lambda = C_d \mathrm{A}_{leak}\sqrt{2g}$ and $\lambda \geq 0$.

6.3 The Model Modification of the Pipeline with Leakage

A leakage flow at some coordinate x_l in the pipeline (Fig. 6.2), denoted by Q_l, is approximately proportional to the square root of the pressure head at this coordinate, H_l, according to:

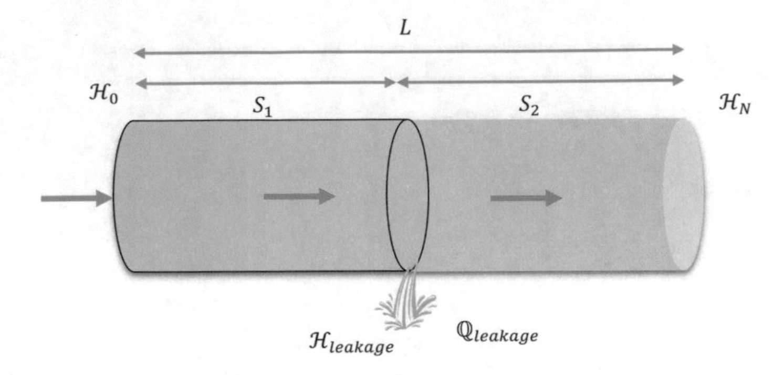

Fig. 6.2 Schematic of pipeline with leakage

$$\mathbb{Q}_l(t) = \sqrt{\lambda \mathcal{H}_l(t)} \tag{6.6}$$

where λ denotes a constant that includes properties such as the cross-sectional area of the leak and a discharge coefficient and $\mathbb{Q}_l$ is the flow through the leak and H_l is pressure head at the leak point [65].

Each leak produces a discontinuity in the mass flow rate. The mass conservation at $\mathbb{Q}_l$ requires that:

It has been assumed that the inlet pressure h_1 and the outlet pressure h_3 are known, which are externally defined by the pump power. The pressure h_2 at the leak point as well as the inlet and outlet flow rates ($\mathbb{Q}_1$ and $\mathbb{Q}_2$) is taken to be the indistinct dynamic variable. According to the continuity equation in pipeline we have,

$$\mathbb{Q}_l = \mathbb{Q}_0 - \mathbb{Q}_N \tag{6.7}$$

where $\mathbb{Q}_0$ and $\mathbb{Q}_N$ are the flows in the beginning section of the pipeline and end of pipeline that leak respectively.

Finally, for the pipeline we can write continuity equation as follow:

$$\frac{\partial \mathcal{H}}{\partial t} + \frac{a^2}{g\text{Å}} \frac{\partial \mathbb{Q}}{\partial x} = 0 \tag{6.8}$$

And the momentum equation as

$$\frac{\partial \mathbb{Q}}{\partial t} + \text{Å} g \frac{\partial}{\partial x} \mathcal{H} + \frac{⅃\mathbb{Q}^2}{ð\text{Å}} = 0 \tag{6.9}$$

Now for the leak detection, the idea is to assume that one leak may occur at each section end, and include its influence in the above model via Eqs. (6.8) and (6.9), with unknown magnitude. By designing an observer to reconstruct the magnitude

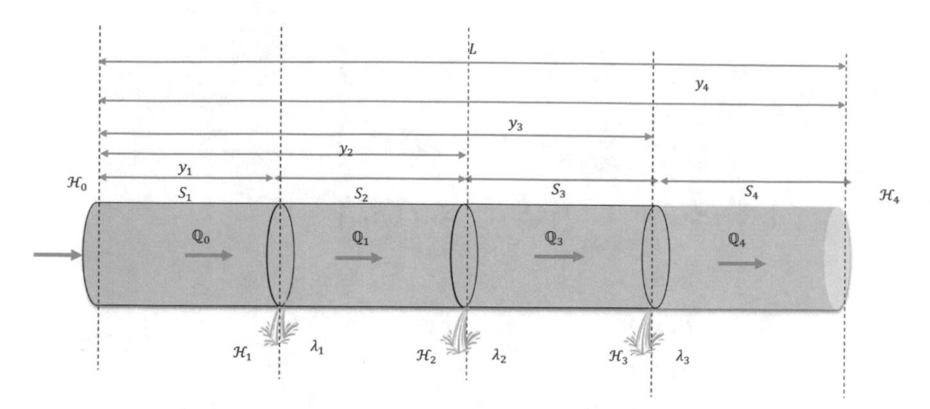

Fig. 6.3 Schematic of pipeline with three leakage

for each leak of this kind, one can detect and isolate them, provided that they indeed occur close to the points a priori considered.

In order to be even more precise, one can further include the location of such possible leaks within the set of unknown variables to be reconstructed by the observer. Leak detection and isolation is then turned to a matter of amplitude and location estimations, and clearly this can in principle be done for any a priori given number of leaks to be detected.

Obviously larger the number of leaks is, the larger the excitation needed for the observer will be (Fig. 6.3).

In the above case of four-section model, one can consider three leaks along the pipeline, located at $y_{1leak} := s_1$ and $y_{2leak} := s_1 + s_2$ and $y_{3leak} := s_1 + s_2 + s_3$ respectively, with y_1 and y_2 and y_3 being unknown.

Of course if L denotes the total length of the pipeline, x_4 will be given by $s_4 = L - s_1 - s_2 - s_3$.

Assuming that the unknown amount (represented by coefficients λ_1, λ_2 and λ_3 in (6.6) are constants, as well as the unknown locations (represented by with s_1, s_2 and s_3 via y_{1leak}, y_{2leak} and y_{3leak}).

Equation (6.6) can be modified by including six additional state variables, $x_8 := s_1, x_9 := s_2, x_{10} := s_3, x_{11} := \lambda_1, x_{12} := \lambda_2, x_{13} := \lambda_3$ as follows:

$$\dot{\mathbb{Q}}_1 = \frac{-\text{Å}g}{s_1}(\mathcal{H}_2 - \mathcal{H}_1) - \frac{\text{Ⅎ}}{2\text{ð}}\frac{\mathbb{Q}_1^2}{\text{Å}}$$

$$\dot{\mathbb{Q}}_2 = \frac{-\text{Å}g}{s_2}(\mathcal{H}_3 - \mathcal{H}_2) - \frac{\text{Ⅎ}}{2\text{ð}}\frac{\mathbb{Q}_1^2}{\text{Å}}$$

$$\dot{\mathbb{Q}}_3 = \frac{-\text{Å}g}{s_3}(\mathcal{H}_4 - \mathcal{H}_3) - \frac{\text{Ⅎ}}{2\text{ð}}\frac{\mathbb{Q}_1^2}{\text{Å}}$$

$$\dot{\mathbb{Q}}_4 = \frac{-\text{Å}g}{L - s_1 - s_2 - s_3}(\mathcal{H}_5 - \mathcal{H}_4) - \frac{\text{Ⅎ}}{2\text{ð}}\frac{\mathbb{Q}_1^2}{\text{Å}}$$

$$
\begin{aligned}
\dot{\mathcal{H}}_2 &= \frac{-a^2}{g\mathring{A}s_1}\left(\mathbb{Q}_2 - \mathbb{Q}_1 - \sqrt{\lambda_1 \mathcal{H}_2}\right) \\
\dot{\mathcal{H}}_3 &= \frac{-a^2}{g\mathring{A}s_2}\left(\mathbb{Q}_3 - \mathbb{Q}_2 - \sqrt{\lambda_2 \mathcal{H}_3}\right) \\
\dot{\mathcal{H}}_4 &= \frac{-a^2}{g\mathring{A}s_3}\left(\mathbb{Q}_4 - \mathbb{Q}_3 - \sqrt{\lambda_3 \mathcal{H}_4}\right) \\
\dot{S}_1 &= 0 \\
\dot{S}_2 &= 0 \\
\dot{S}_3 &= 0 \\
\dot{\lambda}_1 &= 0 \\
\dot{\lambda}_2 &= 0 \\
\dot{\lambda}_3 &= 0
\end{aligned}
\tag{6.10}
$$

Again we have to write this equation with element of x and the form of steady state is as follows:

$$
\begin{aligned}
\dot{x}_1 &= \frac{-\mathring{A}g}{x_8}(x_5 - u_1) - \frac{\Finv}{2\eth\mathring{A}}x_1 \\
\dot{x}_2 &= \frac{-\mathring{A}g}{x_9}(x_6 - x_5) - \frac{\Finv}{2\eth\mathring{A}}x_2 \\
\dot{x}_3 &= \frac{-\mathring{A}g}{x_{10}}(x_7 - x_6) - \frac{\Finv}{2\eth\mathring{A}}x_3 \\
\dot{x}_4 &= \frac{-\mathring{A}g}{l - x_8 - x_9 - x_{10}}(u_2 - x_7) - \frac{\Finv}{2\eth\mathring{A}}x_4 \\
\dot{x}_5 &= \frac{-a^2}{g\mathring{A}x_8}\left(x_2 - x_1 - \sqrt{x_{11}x_5}\right) \\
\dot{x}_6 &= \frac{-a^2}{g\mathring{A}x_9}\left(x_3 - x_2 - \sqrt{x_{12}x_6}\right) \\
\dot{x}_7 &= \frac{-a^2}{g\mathring{A}x_{10}}\left(x_4 - x_3 - \sqrt{x_{13}x_7}\right) \\
\dot{x}_8 &= 0 \\
\dot{x}_9 &= 0 \\
\dot{x}_{10} &= 0 \\
\dot{x}_{11} &= 0 \\
\dot{x}_{12} &= 0 \\
\dot{x}_{13} &= 0
\end{aligned}
\tag{6.11}
$$

Again for finding the state-space of the system Eqs. (6.8), (6.9) should be written in other form as flow [54]:

$$\dot{x}(t) = f(x(t) + g(x(t))u(t)) = f(x(t), u(t))$$
$$y(t) = h(x(t)) \tag{6.12}$$

where $x \in \mathbb{R}^n$ is the state, $u \in \mathbb{R}^m$ is the input, $y \in \mathbb{R}^p$ is the output and f, g, h are sufficiently differentiable vectors function.

6.4 Observer Formulation

The problem of observer design naturally arises in a system approach, as soon as one needs some internal information from external (directly available) measurements.

In general indeed, it is clear that one cannot use as many sensors as signals of interest characterizing the system behavior (for cost reasons, technological constraints, etc.) [66].

In control theory, a state observer is a system that provides an estimate of the internal state of a given real system, from measurements of the input and output of the real system. It is typically computer-implemented, and provides the basis of many practical applications.

Knowing the system state is necessary to solve many control theory problems. In most practical cases, the physical state of the system cannot be determined by direct observation. Instead, indirect effects of the internal state are observed by way of the system outputs. A simple example is that of vehicles in a tunnel: the rates and velocities at which vehicles enter and leave the tunnel can be observed directly, but the exact state inside the tunnel can only be estimated. If a system is observable, it is possible to fully reconstruct the system state from its output measurements using the state observer (Fig. 6.4).

The idea is to design an observer for the leak detection. The system under consideration will be considered to be described by a state-space representation generally of the following form:

$$\dot{x}(t) = f(x(t), u(t)) = f(x(t)) + g(x(t))u(t)$$
$$y(t) = [h_1(x), h_2(x)]^T = h(x(t)) \tag{6.13}$$

where x denotes the state vector, u denotes the vector of known external inputs and y denotes the vector of measured outputs.

Functions f and h will in general be assumed to be C∞ with respect to their arguments, and input functions u(.) to be locally essentially bounded.

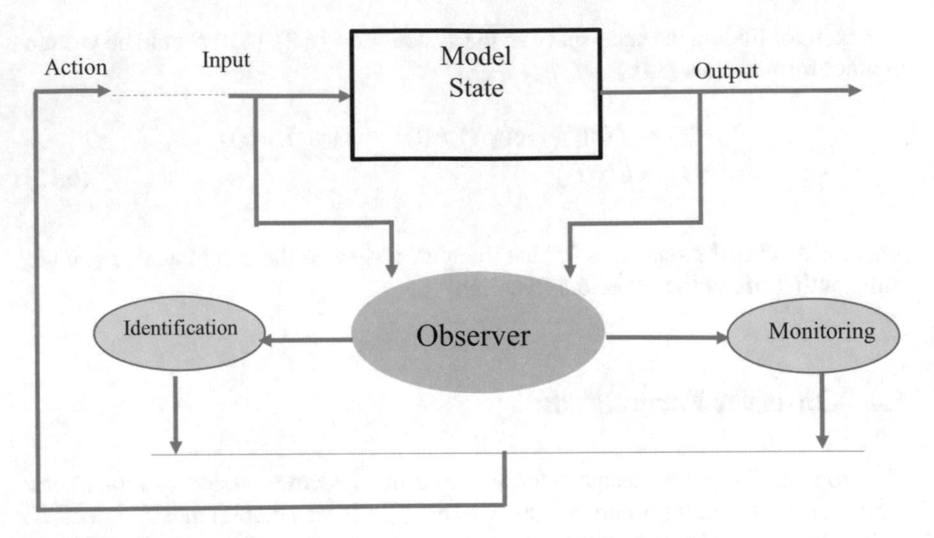

Fig. 6.4 Observer model

6.5 Luenberger Observer

6.5.1 Linear Approaches

Equations (6.8) and (6.9) resulting in (6.13) is in general nonlinear, and a very first approach could be to consider an approximate linearization around some fixed operation regime, giving rise to a standard linear time-invariant system of the form

$$\begin{aligned} \dot{x}(t) &= Ax(t) + Bu(t) \\ y(t) &= Cx(t) \end{aligned} \tag{6.14}$$

Under this form, it is well known that a simple observability condition on A, C guarantees that a so-called Luenberger observer [67–69] can solved for the state reconstruction issue, such as

$$\dot{\hat{x}}(t) = A\hat{x}(t) - K\big(C\hat{x}(t) - y(t)\big) + Bu(t) \tag{6.15}$$

for any matrix K such that $A - KC$ is Hurwitz (i.e., has all its eigenvalues with strictly negative real parts).

In fact, because of the great sensitivity of the operation point with respect to faults like leaks or obstructions, this approach is not successful in general [70].

6.5.2 Nonlinear Approaches Luenberger Extension

The most famous extension of the Luenberger observer is the so-called high-gain observer [71] which applies to systems when it is rewritten under the form

$$\begin{aligned}\dot{\xi}(t) &= A_0\xi(t) + \varphi(\xi(t), u(t)) \\ y(t) &= C_0\xi(t)\end{aligned} \tag{6.16}$$

where $A_0 = \begin{bmatrix} 0 & I_p & & 0 \\ & \ddots & \ddots & \\ & & & I_p \\ 0 & & \cdots & 0 \end{bmatrix}$, $C_0 = \left(I_p 0 \cdots 0\right)$ and $\varphi(\xi, u)$ satisfies the Lipschitz condition in ξ uniformly in u and can be described by

$$\varphi(\xi, u) = \begin{bmatrix} \varphi_1(\xi_1, u) \\ \varphi_2(\xi_1, \xi_2, u) \\ \vdots \\ \varphi_{q-1}\left(\xi_1, \ldots, \xi_{q-1}, u\right) \\ \varphi_q(\xi, u) \end{bmatrix} \text{ for } \xi = \begin{bmatrix} \xi_1 \\ \xi_2 \\ \\ \vdots \\ \xi_q \end{bmatrix} \tag{6.17}$$

with $\xi_i \in R_p$, if $y \in Rp$, and I_p stands for the p × p identity matrix.

This is possible under the uniform observability property [72] and an observer then takes the form

$$\dot{\hat{\xi}}(t) = A_0\hat{\xi}(t) + \varphi\left(\hat{\xi}(t), u(t)\right) - K_0\left[C_0\hat{\xi}(t) - y(t)\right] \tag{6.18}$$

where K_0 is such that $A_0 - K_0C_0$ is Hurwitz, and λ is to be chosen large enough to guaranty convergence, simultaneously allowing the tuning of the convergence rate.

6.6 Lie Derivative

In differential geometry, the Lie derivative named after Sophus Lie, evaluates the change of a tensor field, along the flow of another vector field. This change is coordinate invariant and therefore the Lie derivative [73] is defined on any differentiable manifold.

We now introduce a Lie derivative, which is virtually a directional derivative for a scalar field $\lambda(x)$ with $x \in \mathbb{R}^n$ along the direction of an n-dimensional vector field $f(x)$. The mathematical expression is given as

$$L_f\lambda(x) = \frac{\partial\lambda(x)}{\partial x} f(x) \tag{6.19}$$

Since $\frac{\partial\lambda(x)}{\partial x}x$ is a $1 \times n$ gradient vector of the scalar $\lambda(x)$ and the norm of a gradient vector represents the maximum rate of function value changes, the product of the gradient and the vector field $f(x)$ in system becomes the directional derivative of $\lambda(x)$ along $f(x)$. Therefore, the Lie derivative of a scalar field defined by (6.19) is also a scalar field [74].

If each component of a vector field $h(x) \in \mathbb{R}^p$, p is considered to take a Lie derivative along $f(x) \in \mathbb{R}^n$, then all components can be acted on concurrently and the result is a vector field that has the same dimension as $h(x)$; its ith element is the Lie derivative of the ith component of $h(x)$. Namely, if $h(x) = \left[h_1(x), \ldots, h_p(x)\right]^T$ and each component $h_1(x),\ i = 1, \ldots, p$ is a scalar field, then the Lie derivative of the vector field $h(x)$ is defined as

$$L_f h(x) = \begin{bmatrix} L_f h_1(x) \\ \vdots \\ L_f h_p(x) \end{bmatrix} \tag{6.20}$$

With the Lie derivative concept, we now define an observation space Ω over $\mathbb{R}^n$ as

$$\Omega = \text{span}\left[\text{h(x)}, L_f h(x), \ldots, L_f^{n-1} h(x)\right] \tag{6.21}$$

In other words, this space is spanned by all up to order n Lie derivatives of the output function $h(x)$.

The system is observable if and only if $\dim(d\Omega) = n$. For a proof. See [75].

6.7 Example (Model for Pipe with Two Sections)

Let us recall how such an approach can be applied to leak position and magnitude estimation by considering the model (6.8) (6.9) reduced to two sections, that is $\mathbb{Q}_1$, $\mathcal{H}_2$, $\mathbb{Q}_2$ with $\mathbb{Q}_1$ and $\mathbb{Q}_2$ measured variables and as state variables x_1, x_2, x_3. If this state is extended with a single leak at unknown position z (defined as state variable x_4) and the unknown coefficient λ (defined as state variable x_5), and subject to inputs $u_1 = \mathcal{H}_1$, $u_2 = \mathcal{H}_3$, we have state-space representation under (6.8) and (6.9) and (6.13) as follows:

$$\begin{aligned} \dot{x}(t) &= f(x(t)) + g(x(t))u \\ y &= h(x(t)) \end{aligned} \tag{6.22}$$

With $x := [\mathbb{Q}_1 \mathcal{H}_2 \mathbb{Q}_2 z \lambda]^T = [x_1 x_2 x_3 x_4 x_5]^T$.

Notice that under constant down stream pressure operation, as this will be considered $H_{out} = H_3$.

$$\begin{bmatrix} \dot{x}_1 \\ \dot{x}_2 \\ \dot{x}_3 \\ \dot{x}_4 \\ \dot{x}_5 \end{bmatrix} = \begin{bmatrix} \dot{\mathbb{Q}}_1 \\ \dot{\mathcal{H}}_2 \\ \dot{\mathbb{Q}}_2 \\ \dot{z}_1 \\ \dot{\lambda} \end{bmatrix} = \begin{bmatrix} \frac{-g\AA}{z}(\mathcal{H}_2 - \mathcal{H}_1) \\ -\frac{b^2}{g\AA z}\left(\mathbb{Q}_2 - \mathbb{Q}_1 + \lambda\sqrt{\mathcal{H}_2}\right) \\ -\frac{gA}{L-z}(\mathcal{H}_3 - \mathcal{H}_2) \\ 0 \\ 0 \end{bmatrix}$$

$$= \begin{bmatrix} \frac{-g\AA}{x_4}(x_2 - u_1) \\ -\frac{b^2}{g\AA x_4}\left(x_3 - x_1 + x_4\sqrt{x_2}\right) \\ -\frac{g\AA}{L-x_4}(u_2 - x_2) \\ 0 \\ 0 \end{bmatrix} \tag{6.23}$$

Let the vector of output derivatives $H(x)$, be defined as follows:

$$\hat{H}(x) = \begin{pmatrix} h_1(x) \\ L_f h_1(x) \\ L_f^2 h_1(x) \\ h_2(x) \\ L_f h_2(x) \end{pmatrix} \tag{6.24}$$

where $L^i_f h_j(x)$ represents the ith Lie derivate of $h_j(x)$ in the f vector field direction. For system (6.22), the main objective is the estimation of the state x based on the output measurement y, its derivatives $\hat{H}$(x) and state transformation from output derivatives space, ψ to the original space state, x, defined precisely by $\psi = \hat{H}(x)$.

From Eq. (6.17), the vector H(x) for the system is defined

$$\hat{H}(x) = \begin{pmatrix} \xi_1 \\ \dot{\xi}_1 \\ \ddot{\xi}_1 \\ \xi_2 \\ \dot{\xi}_2 \end{pmatrix} = \begin{pmatrix} h_1(x) \\ L_f h_1(x) \\ L_f^2 h_1(x) \\ h_2(x) \\ L_f h_2(x) \end{pmatrix} = \begin{pmatrix} x_1 \\ -\frac{g\AA}{x_4}(x_2 - u_1) \\ \frac{b^2}{x_4^2}\left(x_3 - x_1 + x_5\sqrt{x_2} - \dot{u}_1\right) \\ x_3 \\ \frac{g\AA}{L-x_4}(u_2 - x_2) \end{pmatrix} \tag{6.25}$$

From (6.16) we can write observer system as follow

$$\dot{x}_1 = x_1$$

$$\hat{\dot{x}}_1 = \frac{-g\AA}{\hat{x}_4}(\hat{x}_2 - u_1)$$

$$
\begin{aligned}
\dot{\hat{x}}_1 &= \frac{b^2}{\hat{x}_4^2}\left(\hat{x}_3 - \hat{x}_1 - \hat{x}_5\sqrt{\hat{x}_2} - \dot{u}_1\right)\\
\hat{x}_2 &= \left(\frac{L - \hat{x}_4}{g\text{Å}}\right)\dot{\hat{x}}_3 + u_2\\
\hat{x}_3 &= x_3\\
\dot{\hat{x}}_3 &= \frac{g\text{Å}}{L - \hat{x}_4}(u_4 - \hat{x}_2)\\
\hat{x}_4 &= \frac{L\dot{\hat{x}}_3 - g\text{Å}(u_1 - u_2)}{\dot{\hat{x}}_3 - \dot{\hat{x}}_1}\\
\hat{x}_5 &= \frac{1}{\sqrt{\hat{x}_2}}\left[\frac{\hat{x}_4}{b^2}\dot{\hat{x}}_1\right] + \frac{1}{\sqrt{\hat{x}_2}}[x_1 - x_3 + \dot{u}_1]
\end{aligned}
\tag{6.26}
$$

6.8 Simulation

In this chapter, we present simulation results in order to evaluate the performance of designed leak detection and isolation scheme. The simulator has the same structure as the model system (6.11).

For those simulations, a pipeline was considered with similar characteristics as summarized in Table 6.1.

For the sake of illustration, here are reported the results for one example of single leak detection and another one for the case of three simultaneous leaks.

In all cases, the initial pressure head at upstream and downstream point of the pipeline were fixed in 16 (m) and 7 (m), respectively. That $u_2 = 2\,\text{m}$, simple signal and $u_1 = 16\,\text{m}$. The simulator was initialized in steady state condition as follows: $\mathbb{Q}_1(0) = \mathbb{Q}_2(0) = 8.3 \times 10^{-3}\left(\frac{\text{m}^3}{\text{s}}\right)$ since, in a free-leak-pipeline, the inflow is equal to out flow; for one leak $z_1(0) = 0.5L$ (m) the discrete point is located in the middle of pipe length, $\lambda(0) = 0\left(\text{m}^{\frac{5}{2}}/\text{s}\right)$ since, at the beginning of the simulation, the pipe is not leaking.

Figure 6.5 presents the simulated pressure head at inlet ($\mathcal{H}_{in} = \mathcal{H}_1$) and outlet ($\mathcal{H}_{out}$) of the pipe.

Table 6.1 Pipeline characteristics

L	220
D	0.220
Å	0.038
b	1250 m/s
g	9.81 m/s^2
f	0

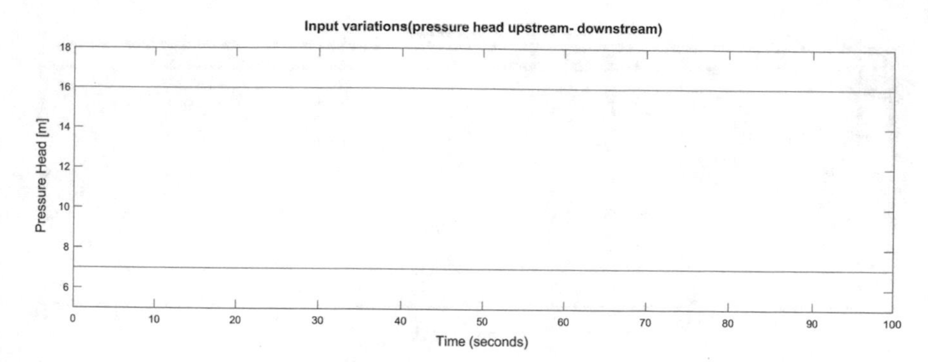

Fig. 6.5 Pressure head at inlet and outlet of pipeline with leakage

Figure 6.6 shows the evolution of the inflow and outflow, $\mathbb{Q}_1$ and $\mathbb{Q}_2$ respectively. That shows after a few second the $\mathbb{Q}_1$ from initial amount $\mathbb{Q}(0)$ is reaching to real amount.

In Fig. 6.7, the parameter valued λ are shown.

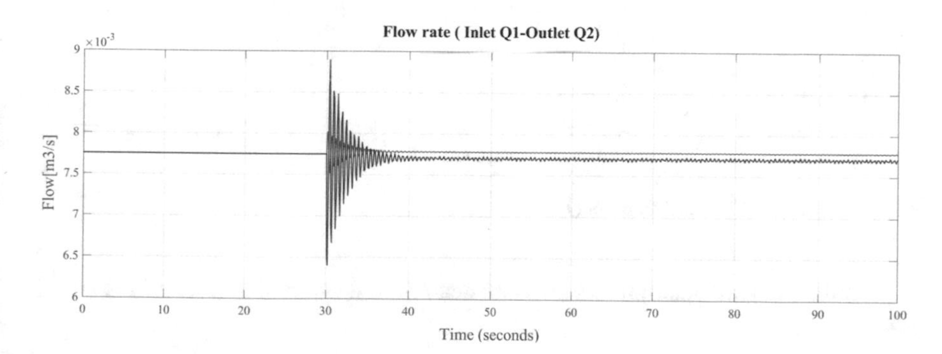

Fig. 6.6 Flow rate at inlet and outlet of pipeline with leakage

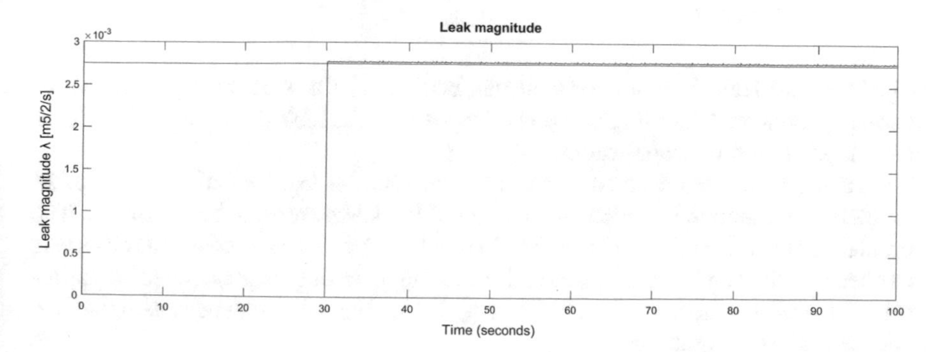

Fig. 6.7 Leak magnitude

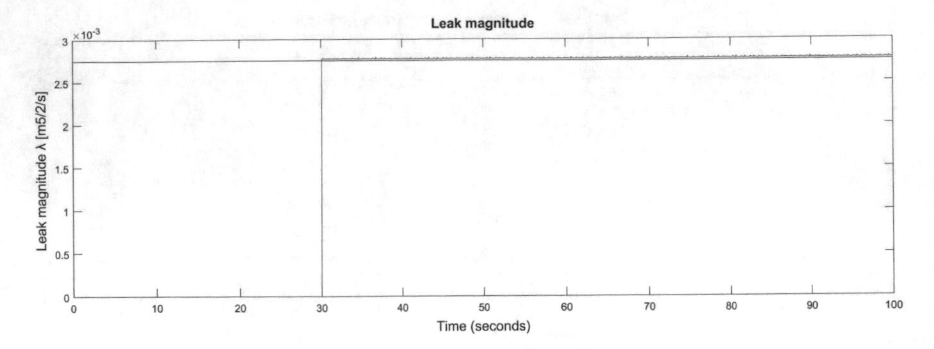

Fig. 6.8 Leak position

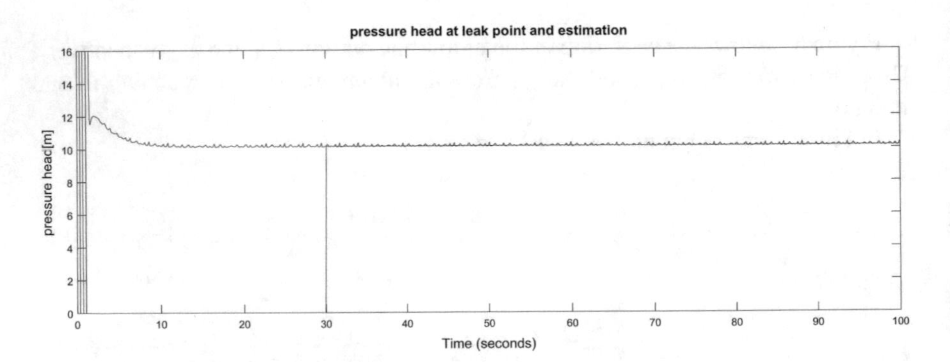

Fig. 6.9 Pressure head at leak point ($\mathcal{H}_2$)

Finally, the leak position and pressure head is well estimated as seen in Figs. 6.8 and 6.9.

6.9 Conclusion

Pipelines and leak detection systems can be found in a wide variety of areas for various products. Accordingly, the challenges that the leak detection system faces vary depending on the application.

In this chapter is presented a model to confront the leak detection and location problem in a pipeline in which the position of the leaks are assumed unknown. The problem of leak detection and isolation in pipelines by means of nonlinear observers has been studied and tested in simulation. The detectability is guaranteed by appropriate excitation, chosen as a small varying input signal, via extensive simulations, and the results are quite promising.

It can be noticed that in the three-leak case which has been considered, the estimation performances could be improved by taking advantage of a first-step study on some equivalent single-leak location allowing to reduce the estimation problem.

But the fairly good results here obtained when directly facing the complete estimation problem are encouraging regarding its possible extension to detection of more leaks.

References

1. Spenser, B., Benichou, N., Bash, A.: Paravalvular leak detection, sealing, and prevention. Google Patents (2007)
2. Billmann, L., Isermann, R.: Leak detection methods for pipelines. Automatica **23**(3), 381–385 (1987)
3. Vítkovský, J.P., Simpson, A.R., Lambert, M.F.: Leak detection and calibration using transients and genetic algorithms. J. Water Resour. Plan. Manag. **126**(4), 262–265 (2000)
4. Murvay, P.S., Silea, I.: A survey on gas leak detection and localization techniques. J. Loss Prev. Process Ind. **25**(6), 966–973 (2012)
5. Fukushima, K., Maeshima, R., Kinoshita, A., Shiraishi, H., Koshijima, I.: Gas pipeline leak detection system using the online simulation method. Comput. Chem. Eng. **24**(2–7), 453–456 (2000)
6. Zhang, J.: Designing a cost-effective and reliable pipeline leak-detection system. Pipes Pipelines Int. **42**(1), 20–26 (1997)
7. Mears, M.N.: Real world applications of pipeline leak detection. In: Pipeline Infrastructure II, pp. 189–209. ASCE (1993)
8. Liou, C.P.: Pipeline leak detection by impulse response extraction (1998)
9. Boaz, L., Kaijage, S., Sinde, R.: An overview of pipeline leak detection and location systems. In: Proceedings of the 2nd Pan African International Conference on Science, Computing and Telecommunications (PACT 2014), pp. 133–137. IEEE (2014)
10. McAtamney, D.E.: Pipeline leak detection system. Google Patents (1994)
11. Covington, M.T.: Liquid pipeline leak detection. Google Patents (1982)
12. Kruka, V.R., Patterson, R.W., Haws, J.H.: Subsea pipeline leak detection. Google Patents (1991)
13. Butts, E.O.: Pipeline leak detection device. Google Patents (1971)
14. Yu, W., Jafari, R.: Modeling and Control of Uncertain Nonlinear Systems with Fuzzy Equations and Z-Number. Wiley (2019)
15. Razvarz, S., Jafari, R.: ICA and ANN modeling for photocatalytic removal of pollution in wastewater. Math. Comput. Appl. **22**(3), 38 (2017)
16. Jafari, R., Razvarz, S., Gegov, A.: Neural network approach to solving fuzzy nonlinear equations using Z-numbers. IEEE Trans. Fuzzy Syst. (2019)
17. Razvarz, S., Jafari, R.: Intelligent techniques for photocatalytic removal of pollution in wastewater. J. Electr. Eng. **5**(1), 321–328 (2017)
18. Jafari, R., Razvarz, S., Gegov, A., Paul, S., Keshtkar, S.: Fuzzy Sumudu transform approach to solving fuzzy differential equations with Z-numbers. In: Advanced Fuzzy Logic Approaches in Engineering Science, pp. 18–48. IGI Global (2019)
19. Jafari, R., Razvarz, S., Gegov, A.: A novel technique to solve fully fuzzy nonlinear matrix equations. In: International Conference on Theory and Applications of Fuzzy Systems and Soft Computing, pp. 886–892. Springer (2018)
20. Jafari, R., Razvarz, S., Gegov, A.: Fuzzy differential equations for modeling and control of fuzzy systems. In: International Conference on Theory and Applications of Fuzzy Systems and Soft Computing, pp. 732–740. Springer (2018)

21. Jafari, R., Yu, W., Razvarz, S., Gegov, A.: Numerical methods for solving fuzzy equations: a survey. Fuzzy Sets Syst. (2019)
22. Jafari, R., Razvarz, S., Gegov, A.: A new computational method for solving fully fuzzy nonlinear systems. In: International Conference on Computational Collective Intelligence, pp. 503–512. Springer (2018)
23. Jafari, R., Razvarz, S., Gegov, A., Paul, S.: Modeling and control of uncertain nonlinear systems. In: 2018 International Conference on Intelligent Systems (IS), pp. 168–173. IEEE (2018)
24. Jafari, R., Razvarz, S., Gegov, A.: A novel technique for solving fully fuzzy nonlinear systems based on neural networks. Vietnam J. Comput. Sci. **7**(1), 93–107 (2020)
25. Razvarz, S., Hernández-Rodríguez, F., Jafari, R., Gegov, A.: Foundation of Z-numbers and engineering applications. In: Latin American Symposium on Industrial and Robotic Systems, pp. 15–24. Springer (2019)
26. Jafari, R., Contreras, M.A., Yu, W., Gegov, A.: Applications of fuzzy logic, artificial neural network and neuro-fuzzy in industrial engineering. In: Latin American Symposium on Industrial and Robotic Systems, pp. 9–14. Springer (2019)
27. Jafari, R., Razvarz, S., Gegov, A., Yu, W.: Fuzzy control of uncertain nonlinear systems with numerical techniques: a survey. In: UK Workshop on Computational Intelligence, pp. 3–14. Springer (2019)
28. Jafari, R., Razvarz, S., Yu, W., Gegov, A., Goodwin, M., Adda, M.: Genetic algorithm modeling for photocatalytic elimination of impurity in wastewater. In: Proceedings of SAI Intelligent Systems Conference, pp. 228–236. Springer (2019)
29. Tatchum, M., Gegov, A., Jafari, R., Razvarz, S.: Parallel distributed compensation for voltage controlled active magnetic bearing system using integral fuzzy model. In: 2018 International Conference on Intelligent Systems (IS), pp. 190–198. IEEE (2018)
30. Razvarz, S., Jafari, R., Gegov, A.: Solving partial differential equations with Bernstein neural networks. In: UK Workshop on Computational Intelligence, pp. 57–70. Springer (2018)
31. Jafarian, A., Jafari, R.: New iterative approach for solving fully fuzzy polynomials. Int. J. Fuzzy Math. Syst. **3**(2), 75–83
32. Jafarian, A., Jafari, R.: New method for solving fuzzy polynomials. Adv. Fuzzy Math. **8**(1), 25–33 (2013)
33. Jafarian, A., Jafari, R.: An iterative method for solving fuzzy polynomials by fuzzy neural networks (2012)
34. Jafarian, A., Jafari, R.: (2012) Simulation and evaluation of fuzzy polynomials by feed-back neural networks
35. Jafari, R., Yu, W.: Fuzzy control for uncertainty nonlinear systems with dual fuzzy equations. J. Intell. Fuzzy Syst. **29**(3), 1229–1240 (2015)
36. Jafari, R., Yu, W.: Fuzzy modeling for uncertainty nonlinear systems with fuzzy equations. Math. Probl. Eng. (2017)
37. Verde, C., Visairo, N., Gentil, S.: Two leaks isolation in a pipeline by transient response. Adv. Water Resour. **30**(8), 1711–1721 (2007)
38. Verde, C., Visairo, N.: Identificability of multi-leaks in a pipeline. In: Proceedings of the 2004 American Control Conference, pp. 4378–4383. IEEE (2004)
39. Verde, C., Visairo, N.: Multi-leak isolation in a pipeline by unsteady state test. In: 44th IEEE Conference on Decision and Control and European Control Conference ECC, pp. 1711–1721 (2005)
40. Verde, C.: Accommodation of multi-leak location in a pipeline. Control Eng. Pract. **13**(8), 1071–1078 (2005)
41. Carrera, R., Verde, C., Cayetano, R.: A SCADA expansion for leak detection in a pipeline. Sensors **2300**(2320), 2340 (2015)
42. Verde, C.: Leakage location in pipelines by minimal order nonlinear observer. In: Proceedings of the 2001 American Control Conference (Cat. No. 01CH37148), pp. 1733–1738. IEEE (2001)
43. Kowalczuk, Z., Gunawickrama, K.: Leak detection and isolation for transmission pipelines via nonlinear state estimation. IFAC Proc. Vol. **33**(11), 921–926 (2000)
44. Verde, C., Torres, L.: Modeling and Monitoring of Pipelines and Networks. Springer (2017)

45. Antory, D.: Application of a data-driven monitoring technique to diagnose air leaks in an automotive diesel engine: a case study. Mech. Syst. Signal Process. **21**(2), 795–808 (2007)
46. Ballesta, C., Berindoague, R., Cabrera, M., Palau, M., Gonzales, M.: Management of anastomotic leaks after laparoscopic Roux-en-Y gastric bypass. Obes. Surg. **18**(6), 623–630 (2008)
47. Vickers, A.L.: The future of water conservation: challenges ahead. J. Contemp. Water Res. Educ. **114**(1), 8 (1999)
48. Environment Canada: Threats to Water Availability in Canada. National Water Research Institute, Burlington (2004)
49. Al-Dhowalia, K., Shammas, N.K., Quraishi, A., Al-Muttair, F.: Assessment of Leakage in the Riyadh Water Distribution Network. First Progress Report. King Abdulaziz City for Science and Technology (1989)
50. Billmann, L., Isermann, R.: Leak detection methods for pipelines. IFAC Proc. Vol. **17**(2), 1813–1818 (1984)
51. Benkherouf, A., Allidina, A.: Leak detection and location in gas pipelines. IEE Proc. D Control Theory Appl. **2**, 142–148 (1988)
52. Verde, C.: Minimal order nonlinear observer for leak detection. J. Dyn. Syst. Meas. Control **126**(3), 467–472 (2004)
53. Angulo, M.T., Verde, C.: Second-order sliding mode algorithms for the reconstruction of leaks. In: 2013 Conference on Control and Fault-Tolerant Systems (SysTol), pp. 566–571. IEEE (2013)
54. Besançon, G., Georges, D., Begovich, O., Verde, C., Aldana, C.: Direct observer design for leak detection and estimation in pipelines. In: 2007 European Control Conference (ECC), pp. 5666–5670. IEEE (2007)
55. Krass, W., Kittel, A., Uhde, A.: Pipeline technology. Petroleum oil-long-distance pipelines. Pipelinetechnik. Mineralolfernleitungen. (1979)
56. Razvarz, S., Vargas-Jarillo, C., Jafari, R., Gegov, A.: Flow control of fluid in pipelines using PID controller. IEEE Access **7**, 25673–25680 (2019)
57. Razvarz, S., Vargas-Jarillo, C., Jafari, R.: Pipeline monitoring architecture based on observability and controllability analysis. In: 2019 IEEE International Conference on Mechatronics (ICM), 18–20 Mar 2019, pp. 420–423 (2019)
58. Razvarz, S., Jafari, R., Vargas-Jarillo, C., Gegov, A., Forooshani, M.: Leakage detection in pipeline based on second order extended Kalman filter observer. IFAC-PapersOnLine **52**(29), 116–121 (2019). https://doi.org/10.1016/j.ifacol.2019.12.631
59. Razvarz, S., Jafari, R., Vargas-Jarillo, C.: Modelling and analysis of flow rate and pressure head in pipelines. In: 2019 16th International Conference on Electrical Engineering, Computing Science and Automatic Control (CCE), pp. 1–6. IEEE (2019)
60. Jafari, R., Razvarz, S., Vargas-Jarillo, C., Yu, W.: Control of flow rate in pipeline using PID controller. In: 2019 IEEE 16th International Conference on Networking, Sensing and Control (ICNSC), 9–11 May 2019, pp. 293–298 (2019)
61. Jafari, R., Razvarz, S., Vargas-Jarillo, C., Gegov, A.: Blockage detection in pipeline based on the extended Kalman filter observer. Electronics **9**(1), 91–107 (2020)
62. Jafari, R., Razvarz, S., Vargas-Jarillo, C., Gegov, A.E.: The effect of baffles on heat transfer. In: ICINCO (2). pp. 607–612 (2019)
63. Çengel, Y.A., Turner, R.H., Cimbala, J.M., Kanoglu, M.: Fundamentals of Thermal-Fluid Sciences, vol. 703. McGraw-Hill, New York (2001)
64. Saleta, M.E., Tobia, D., Gil, S.: Experimental study of Bernoulli's equation with losses. Am. J. Phys. **73**(7), 598–602 (2005)
65. Houghtalen, R.J., Osman, A., Hwang, N.H.: Fundamentals of Hydraulic Engineering Systems. Prentice Hall, New York (2016)
66. Besançon, G.: Nonlinear Observers and Applications, vol. 363. Springer (2007)
67. Zeitz, M.: The extended Luenberger observer for nonlinear systems. Syst. Control Lett. **9**(2), 149–156 (1987)

68. Ciccarella, G., Dalla Mora, M., Germani, A.: A Luenberger-like observer for nonlinear systems. Int. J. Control **57**(3), 537–556 (1993)
69. Birk, J., Zeitz, M.: Extended Luenberger observer for non-linear multivariable systems. Int. J. Control **47**(6), 1823–1836 (1988)
70. Guillén, M., Dulhoste, J.F., Besancon, G., Rubio, I., Santos, R., Georges, D.: Leak detection and location based on improved pipe model and nonlinear observer. In: 2014 European Control Conference (ECC), pp. 958–963. IEEE (2014)
71. Gauthier, J.P., Hammouri, H., Othman, S.: A simple observer for nonlinear systems applications to bioreactors. IEEE Trans. Autom. Control **37**(6), 875–880 (1992)
72. Gauthier, J., Bornard, G.: Observability for any u(t) of a class of nonlinear systems. IEEE Trans. Autom. Control **26**(4), 922–926 (1981)
73. Yano, K.: The Theory of Lie Derivatives and Its Applications. Courier Dover Publications (2020)
74. Kocarev, L., Parlitz, U., Hu, B.: Lie derivatives and dynamical systems. Chaos Solitons Fract. **9**(8), 1359–1366 (1998)
75. Nijmeijer, H., Van der Schaft, A.: Nonlinear Dynamical Control Systems, vol. 175. Springer (1990)

Chapter 7
Blockage Detection in Pipeline

7.1 Introduction

Along with the development of industrial requirements, more and more long-distance pipelines have built and served. According to the experiences, pipelines, as a means of transport, are the safest but this does not mean they are risk-free. When passing through the harsh environment, pipelines may be broken or blocked which would cause huge economic loss. Therefore, assuring the reliability of the pipeline infrastructure has become a critical need for the energy sector. Nowadays, by introducing neural network, as new method [1–23] for monitoring the pipelines tries to found flaw in pipelines.

Leaks, as well as partial or complete blockages, can be considered as a usual fault arising in pipelines that creates troubles [24–26]. Leaks result in loss of fluid that causes a loss in pressure, efficiency as well as economic expense, also on occasions it may impact the environment. Blockage barricade flow, which causes a loss in pressure, and so enhances the required pumping expenses in order to dominate the loss in pressure, also occasionally blockage results in absolute termination of functioning [27, 28]. Advanced diagnosis, as well as an exact place of leaks or blockages, can help in avoiding the troubles created by such faults and also develop the correct timing decision in order to deal with faults for preventing or diminishing production/performance suspension [29–33].

Blockages have been classified based on their physical extent related to the entire length of the system. Localized contractions, which are taken to be point discontinuities, are defined as discrete blockages. Blockages that contain remarkable length related to the entire pipe length are defined as extended blockages [34]. Dissimilar to the leaks inside piping systems, blockages do not produce obvious exterior indicators for their location to be mentioned as the release and agglomeration of fluids surrounding the pipe. Oftentimes intrusive approaches, utilizing devices the insertion of a closed-circuit camera or a robotic PIG (pipeline intervention gadget) [35–38], are needed in order to specify the place of blockages. Insertion of a camera or robotic pig can lead to several uncertainties related to the speed of travel and the

S. Razvarz et al., *Flow Modelling and Control in Pipeline Systems*, Studies in Systems, Decision and Control 321, https://doi.org/10.1007/978-3-030-59246-2_7

distance of travel in the midst of the insertion point and the blockage such that on many occasions cases the camera or the robotic pig is trapped into the blockage and leads to greater trouble in this case compared with the appearance of the blockage solely [39, 40].

Recently, flow analysis is developed on the basis of efficient methods in order to identify blockages. The developed techniques utilize fluid transients according to the responding of the system to an injected transient in order to detect, locate, as well as to measure blockages, which have demonstrated a promising development. In [41, 42] a time reflection technique is suggested in order to detect partial blockage of discrete and extended types in a single pipeline for different numbers of blockages. In [43] an impulse response technique is proposed in order to detect leaks as well as partial blockages of discrete type in a single pipeline, and also the technique is tested numerically. In [44] the authors used the damping of fluid transients on the basis of the analytical solution as well as experimental verifications in order to detect partial single discrete blockage in a single pipeline. In [45] the numerical outcomes with laboratory experiments are compared and concluded that the blockage location can be identified with nearly no error, whilst the size identification contains some errors.

The most current published researches are based on designing the observer, controller or fault diagnosis techniques [46–53]. Nevertheless, investigating the observability as well as controllability of the pipeline system is exceptionally significant. The classical techniques for determining the observability as well as controllability is in direct relation with the fullness of the rank of the discrimination matrices. In this chapter, a novel method is proposed in order to analyze one-dimensional modeling of transients in a pipeline, generally utilized for identification and place of blockage using model-based techniques. The modeling process is based on discretization with the finite-difference technique of classical mass as well as continuity equations. The discretization results in a system of ordinary differential equations with boundary conditions, which demonstrate faults as well as pipeline accessories. The states of the produced system are flows, pressure heads, the blockage and a parameter concerned with the blockage intensity.

7.2 Blockage Modelling

In this section, the convective change in velocity, as well as the compressibility in the line of length (L), are neglected, and the liquid density (ρ), the flow rate $(\mathbb{Q})$, and pressure $(\wp)$ at the inlet and outlet of the pipeline are measurable for evaluation. The cross-sectional area $(\mathring{A})$ of the pipe is constant throughout the pipe. The designed pipeline is shown in Fig. 7.1.

The dynamics of fluid in a pipeline is defined on the basis of the mass and momentum, and the conservation equations in hydrodynamics. In order to obtain the mass and momentum in hydrodynamics by implementing Newton's second law $(F = ma)$ to a control volume in the continuum and body force pipe [54, 55] $(s = \frac{\dashv}{2\eth} \upsilon)$ the following relation is obtained,

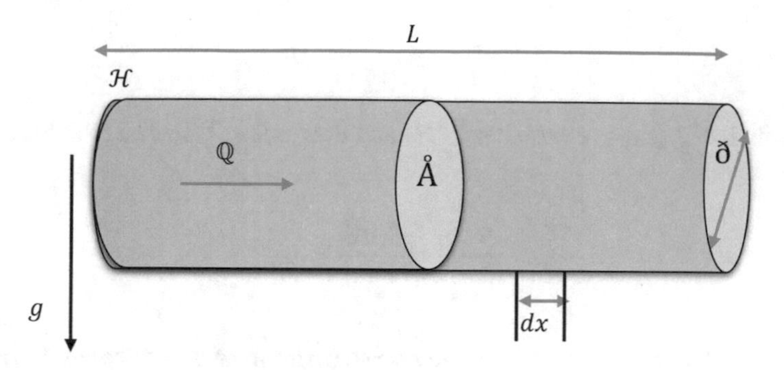

Fig. 7.1 Schematic of pipeline

$$\frac{\partial v}{\partial t}+\frac{1}{\rho}\frac{\partial \wp}{\partial x}+\frac{ⅎ}{2ð}v=0 \tag{7.1}$$

By substituting $v=\frac{ℚ}{Å}$ and $p=\rho g\mathcal{H}$ in (7.1) the following relation is extracted,

$$\frac{\partial ℚ}{\partial t}+Åg\frac{\partial}{\partial x}\mathcal{H}+\frac{ⅎ}{2ð}\frac{ℚ_1^2}{Å}=0 \tag{7.2}$$

where $\mathcal{H}$ is the pressure head (m), $ℚ$ is the flow rate (m^3/s), x is the length coordinate (m), t is the time coordinate (s), g is the gravity $\left(m/s^2\right)$, Å is the section area $\left(m^2\right)$, ð is the diameter (m), and f is the friction coefficient.

In most of model coefficient is considered as a constant, even if it is sometimes updated known to depend on the so-called Reynolds number (Re) and the roughness coefficient of the pipe (e). The Swamee-Jain equation [56, 57] describes this coefficient value for a pipe with a circular section of diameter (ð) as:

$$\mathrm{f}=\left(\frac{0.5}{\ln\left[0.27\left(\frac{e}{ð}\right)+5.74\frac{1}{Re^{0.9}}\right]}\right)^2 \tag{7.3}$$

where the Reynolds number can be calculated with:

$$Re=4\frac{\rho ℚ}{\pi ð\mu}=\frac{\rho Vð}{\mu} \tag{7.4}$$

where ρ is the fluid density and μ is the fluid viscosity. Equation (7.3) is valid for $10^{-8}<\frac{e}{ð}<0.01$ and $5000<Re<10^8$.

In order to derive the continuity equation, by applying the law of conservation of mass and using the Reynolds transport theorem to a control volume and simplifying, the following relation is obtained,

$$\frac{\partial p}{\partial t} + \rho a^2 \frac{\partial v}{\partial x} = 0 \tag{7.5}$$

By substituting the pressure head ($\mathcal{H}$) and flow rate (Q) in (7.5) the following is extracted,

$$\frac{\partial \mathcal{H}}{\partial t} + \frac{a^2}{g\text{Å}} \frac{\partial \mathbb{Q}}{\partial x} = 0 \tag{7.6}$$

in which a is the velocity of the pressure wave (m/s) in an elastic conduit filled with a slightly compressible fluid.

The pressure head ($\mathcal{H}$) and flow rate ($\mathbb{Q}$) are functions of position and time as $\mathcal{H}(x, t)$ and $\mathbb{Q}(x, t)$, where $x \in [0, L]$, also L is the length of pipe.

For the system with small changes in the flow rate $\mathbb{Q}$, the momentum equation from the nonlinear system can be linearized as below

$$\frac{\partial \mathbb{Q}}{\partial t} + \text{Å} g \frac{\partial}{\partial x} \mathcal{H} + \frac{\text{Ⅎ}}{2\text{ð}} \frac{\mathbb{Q}_1^2}{\text{Å}} = 0 \tag{7.7}$$

The pipeline model is designed based on (7.6) and (7.7). Obtaining the solutions of these equations are complex. However, several methods are utilized in order to numerically integrate them. Some of the main methods such as characteristics and finite difference techniques are proposed by Wyllie [58, 59]. In this work, the finite difference method is applied as it is a simple way to get a more convenient model in order to observe and control the structure of the nonlinear system. The finite difference method is a discretization technique which divides the entire pipeline into N number of sections [60].

Blockages have been considered as usual faults in pipes as well as pipeline networks. They can be generated because of the agglomeration of the transported fluid or the partial lock of a valve. Blockages decrease the performance of pipeline systems, and also could create crucial harm in the security of the plant. Figure 7.2 presents a blockage, which arises in a pipe.

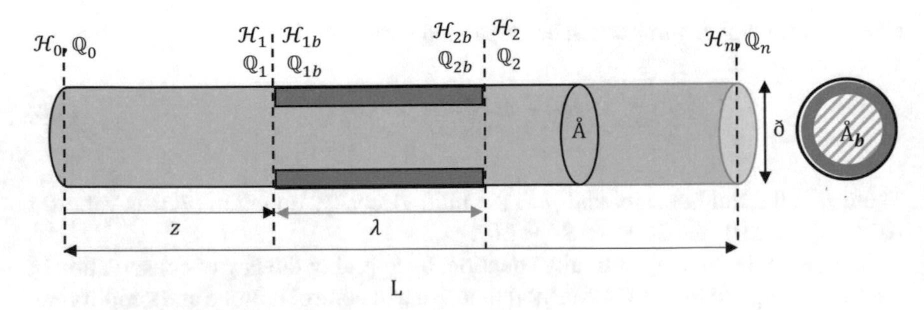

Fig. 7.2 Blockage in the pipeline

A blocked is a reduced cross-sectional area of the pipe with significant length λ. To model the blocked stretch, the lockage area, denoted by $\mathring{A}_b$, is quantified as a percentage of the pipeline area $\mathring{A}$. The subscript b signifies the blockage.

Pressure $\mathcal{H}_1$ varies due to the blockage and this changed pressure is displayed by $\mathcal{H}_{1b}$.

By applying Bernoulli's principle and continuity equations between point 1 (before the blockage) and point 1b (in the blockage), we have [61],

$$\mathcal{H}_1 + \frac{V_1^2}{2g} + Z_1 + h_{w1} = \mathcal{H}_{1b} + \frac{V_{1b}^2}{2g} + Z_{1b} + h_{w1b} + \varepsilon \tag{7.8}$$

where $\mathcal{H}$ denotes the pressure head of the fluid, ν denotes the fluid flow speed, ρ is considered to be the density of the fluid, Z is taken to be the elevation at points blockage, h_w denotes energy inputs with pumps or turbines, and ε is the pressure losses in the pipeline section. Since levels Z_1, as well as Z_{1b}, are equivalent, and also energy inputs with pumps and/or turbines are zero (in this case), so we have

$$\mathcal{H}_1 + \frac{\nu_1^2}{2g} = \mathcal{H}_{1b} + \frac{\nu_{1b}^2}{2g} + \varepsilon \tag{7.9}$$

Based on the continuity of conservation of mass we have

$$\rho\nu_1\mathring{A}_1 = \rho\nu_{1b}\mathring{A}_{1b} \tag{7.10}$$

where $\mathring{A}$ is taken to be the cross-sectional area of the pipe, also $\mathring{A}_b$ is taken to be the cross-sectional area at the blockage. Hence,

$$\nu_{1b} = \frac{\nu_1\mathring{A}}{\mathring{A}_b} \tag{7.11}$$

By substituting Eq. (7.11) in Eq. (7.9) the following relation is extracted

$$\mathcal{H}_{1b} = \mathcal{H}_1 + \frac{\nu_1^2}{2g}\left(1 - \left(\frac{\mathring{A}}{\mathring{A}_b}\right)^2\right) - \varepsilon \tag{7.12}$$

It could be difficult to analytically calculate ε, as it is a function of the flow and the blockage geometry. Hence, for practical intentions, the below-mentioned relation, which contains a discharge coefficient ξ is utilized:

$$\mathcal{H}_{1b} = \mathcal{H}_1 + \xi\frac{\mathring{A}_b^2\nu_1^2 - \mathring{A}^2\nu_1^2}{2g\mathring{A}_b^2} \tag{7.13}$$

which, stated in regards to the volumetric flow $\mathbb{Q} = v\mathring{A}$, becomes

$$\mathcal{H}_{1b} = \mathcal{H}_1 + \xi \frac{\mathring{A}_b^2 \mathbb{Q}_1^2 - \mathring{A}^2 \mathbb{Q}_1^2}{2g\mathring{A}^2 \mathring{A}_b^2} \tag{7.14}$$

This model considers the two circumstances happening at the two edges of the blocked section: a constriction happening upstream of the blockage (from $\mathring{A}$ to $\mathring{A}_b$) and a distension happening downstream (from $\mathring{A}_b$ to $\mathring{A}$).

The following equations describe the dynamics of pressures head and flows before blockage ($\mathcal{H}_1$, $\mathbb{Q}_1$) and after it ($\mathcal{H}_{1b}$, $\mathbb{Q}_{1b}$) respectively. The flow dynamics are then described via the flow rate in each section and the pressures head at the end of each section.

$$\begin{aligned}
\dot{\mathbb{Q}}_0 &= -\mathring{A}g \frac{\mathcal{H}_1 - \mathcal{H}_0}{z} - \frac{\text{Ⅎ}}{2\eth} \frac{\mathbb{Q}_0^2}{\mathring{A}} \\
\dot{\mathbb{Q}}_1 &= -\mathring{A}g \frac{\mathcal{H}_1 - \mathcal{H}_0}{z} - \frac{\text{Ⅎ}}{2\eth} \frac{\mathbb{Q}_1^2}{\mathring{A}} \\
\dot{\mathbb{Q}}_{2b} &= -\mathring{A}_b g \frac{(\mathcal{H}_{2b} - \mathcal{H}_{1b})}{\lambda} - \frac{\text{Ⅎ}_b}{2\eth} \frac{\mathbb{Q}_{2b}^2}{\mathring{A}_b} \\
\dot{\mathbb{Q}}_n &= -\mathring{A}g \frac{(\mathcal{H}_n - \mathcal{H}_2)}{(L - z - \lambda)} - \frac{\text{Ⅎ}}{2\eth} \frac{\mathbb{Q}_n^2}{\mathring{A}} \\
\dot{\mathcal{H}}_0 &= -\frac{a^2}{g\mathring{A}} \frac{\mathbb{Q}_1 - \mathbb{Q}_0}{z} \\
\dot{\mathcal{H}}_1 &= -\frac{a^2}{g\mathring{A}} \frac{\mathbb{Q}_1 - \mathbb{Q}_0}{z} \\
\dot{\mathcal{H}}_{2b} &= -\frac{a^2}{g\mathring{A}_b} \frac{(\mathbb{Q}_{2b} - \mathbb{Q}_{1b})}{\lambda} \\
\dot{\mathcal{H}}_n &= -\frac{a^2}{g\mathring{A}} \frac{(\mathbb{Q}_n - \mathbb{Q}_2)}{(L - z - \lambda)}
\end{aligned} \tag{7.15}$$

where $\mathcal{H}_{1b}$ and $\mathcal{H}_2$ are calculated using Bernoulli's and continuity equation [61, 62] as below

$$\begin{aligned}
\mathcal{H}_{1b} &= \mathcal{H}_1 + \xi \frac{\mathring{A}_b^2 \mathbb{Q}_1^2 - \mathring{A}^2 \mathbb{Q}_1^2}{2g\mathring{A}^2 \mathring{A}_b^2} \\
\mathcal{H}_2 &= \mathcal{H}_{2b} + \xi \frac{\mathring{A}^2 \mathbb{Q}_{2b}^2 - \mathring{A}_b^2 \mathbb{Q}_{2b}^2}{2g\mathring{A}^2 \mathring{A}_b^2}
\end{aligned} \tag{7.16}$$

and at the point of the blockage, the flow rates are equal such that $\mathbb{Q}_1 = \mathbb{Q}_{1b}$ and $\mathbb{Q}_{2b} = \mathbb{Q}_2$. In (7.15) and (7.16), $\mathcal{H}_0, \mathcal{H}_1, \mathcal{H}_{1b}, \mathcal{H}_2, \mathcal{H}_{2b}, \mathcal{H}_n$ are the pressure head variables and $\mathbb{Q}_0, \mathbb{Q}_1, \mathbb{Q}_{1b}, \mathbb{Q}_2, \mathbb{Q}_{2b}, \mathbb{Q}_n$ are the flow rate variables. The idea in this chapter is to design the observer and controller for the systems mentioned in (7.6) and (7.7).

Control variables in the different cases can be presented by the pressures head and flow rates at the beginning and end of the pipe. The solution of the discretized model, using any technique needs determining some boundary conditions that present known values of the variables at the border of the researched area. Boundary conditions may just be pressures head, just flows, or combining both. The selection of boundary conditions alters the construction of the models, also may even change the number of equations wherein the discretized model is subdivided. In real systems, boundary conditions are defined using the elements existing in the hydraulic system. Figure 7.3 represents the possible boundary conditions in a scheme of a discretized pipeline. In Fig. 7.3, we have six cases for the modeling, based on choosing the control variables.

Case 1: The initial and boundary conditions that can be controlled and measured are the pressure heads in the beginning and end of the pipeline, see Fig. 7.3A. The

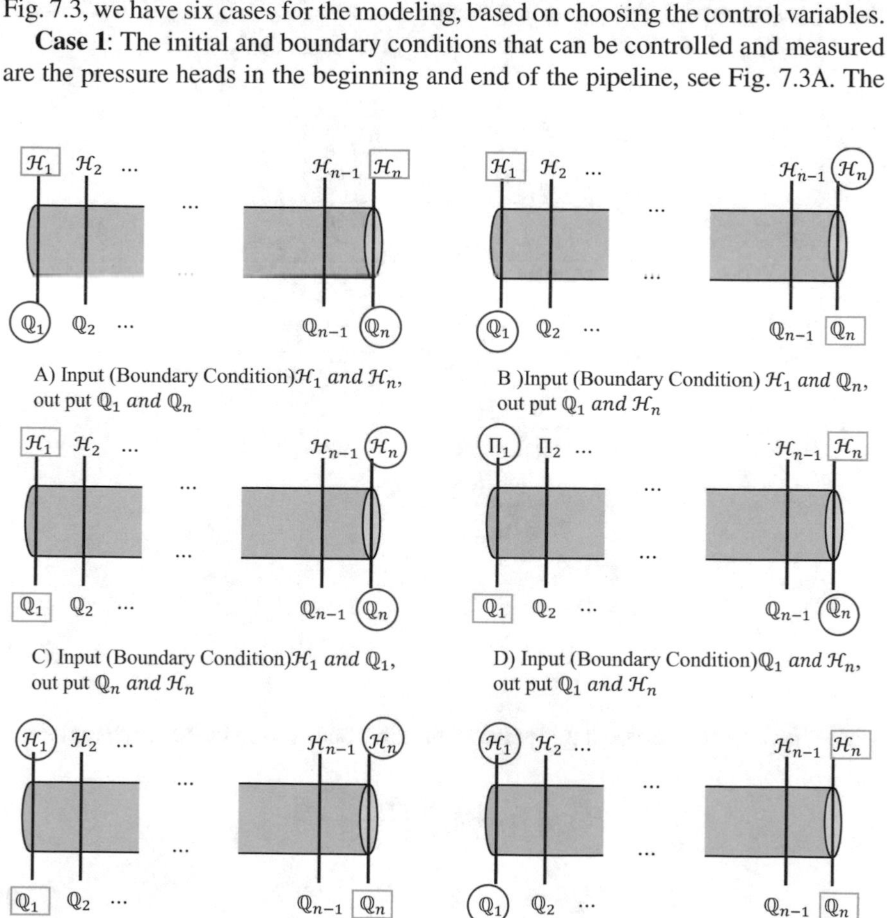

A) Input (Boundary Condition)$\mathcal{H}_1$ *and* $\mathcal{H}_n$, out put $\mathbb{Q}_1$ *and* $\mathbb{Q}_n$

B)Input (Boundary Condition) $\mathcal{H}_1$ *and* $\mathbb{Q}_n$, out put $\mathbb{Q}_1$ *and* $\mathcal{H}_n$

C) Input (Boundary Condition)$\mathcal{H}_1$ *and* $\mathbb{Q}_1$, out put $\mathbb{Q}_n$ *and* $\mathcal{H}_n$

D) Input (Boundary Condition)$\mathbb{Q}_1$ *and* $\mathcal{H}_n$, out put $\mathbb{Q}_1$ *and* $\mathcal{H}_n$

E) Input (Boundary Condition)$\mathbb{Q}_1$ *and* $\mathbb{Q}_n$, out put $\mathcal{H}_1$ *and* $\mathcal{H}_n$

F) Input (Boundary Condition)$\mathcal{H}_n$ *and* $\mathbb{Q}_n$, out put $\mathcal{H}_1$ *and* $\mathbb{Q}_1$

Fig. 7.3 Boundary conditions

input conditions are stated as

$$\begin{cases} \mathcal{H}(0,t) = \mathcal{H}_{in}(t) \\ \mathcal{H}(L,t) = \mathcal{H}_{out}(t) \end{cases} \tag{7.17}$$

and output values are

$$\begin{cases} \mathbb{Q}(0,t) = \mathbb{Q}_{in}(t) \\ \mathbb{Q}(L,t) = \mathbb{Q}_{out}(t) \end{cases} \tag{7.18}$$

The vectors $u(t)$ and $y(t)$ contain the system inputs and outputs respectively. In the suggested model, the inputs and outputs are $x = \left(\mathbb{Q}_0\mathbb{Q}_1\mathbb{Q}_{2b}\mathbb{Q}_n\mathcal{H}_1\mathcal{H}_{2b}\lambda z\mathring{A}_b\right)^T$, $u := [\mathcal{H}_{in}\mathcal{H}_{out}]^T = (\mathcal{H}_0\mathcal{H}_n)^T$ and $y = (\mathbb{Q}_0\mathbb{Q}_n)^T$, respectively. We have,

$$\begin{aligned}
\dot{\mathbb{Q}}_0 &= -\mathring{A}g\frac{\mathcal{H}_1 - \mathcal{H}_0}{z} - \frac{\text{⅃}}{2\eth}\frac{\mathbb{Q}_0^2}{\mathring{A}} \\
\dot{\mathbb{Q}}_1 &= -\mathring{A}g\frac{\mathcal{H}_1 - \mathcal{H}_0}{z} - \frac{\text{⅃}}{2\eth}\frac{\mathbb{Q}_1^2}{\mathring{A}}Q_1 \\
\dot{\mathbb{Q}}_{2b} &= -\mathring{A}_b g\frac{(\mathcal{H}_{2b} - \mathcal{H}_{1b})}{\lambda} - \frac{\text{⅃}_b}{2\eth}\frac{\mathbb{Q}_{2b}^2}{\mathring{A}_b} \\
\dot{\mathbb{Q}}_n &= -\mathring{A}g\frac{(\mathcal{H}_n - \mathcal{H}_2)}{(L - z - \lambda)} - \frac{\text{⅃}}{2\eth}\frac{\mathbb{Q}_n^2}{\mathring{A}} \\
\dot{\mathcal{H}}_1 &= -\frac{a^2}{g\mathring{A}}\frac{\mathbb{Q}_1 - \mathbb{Q}_0}{z} \\
\dot{\mathcal{H}}_{2b} &= -\frac{a^2}{g\mathring{A}_b}\frac{(\mathbb{Q}_{2b} - \mathbb{Q}_{1b})}{\lambda} \\
\dot{\lambda} &= 0 \\
\dot{z} &= 0 \\
\dot{\mathring{A}}_b &= 0
\end{aligned} \tag{7.19}$$

where $\mathcal{H}_{1b}$ and $\mathcal{H}_2$ are calculated using Bernoulli's and continuity equations as below,

$$\begin{aligned}
\mathcal{H}_{1b} &= \mathcal{H}_1 + \xi\frac{\mathring{A}_b^2\mathbb{Q}_1^2 - \mathring{A}^2\mathbb{Q}_1^2}{2gA^2A_b^2} \\
\mathcal{H}_2 &= \mathcal{H}_{2b} + \xi\frac{\mathring{A}^2\mathbb{Q}_{2b}^2 - \mathring{A}_b^2\mathbb{Q}_{2b}^2}{2g\mathring{A}^2\mathring{A}_b^2}
\end{aligned} \tag{7.20}$$

Case 2: The boundary conditions that can be controlled and measured are the flow rate and pressure head in the beginning and end of the pipeline, respectively, see Fig. 7.3B. The input conditions are stated as

$$\begin{cases} \mathbb{Q}(0,t) = \mathbb{Q}_{in}(t) \\ \mathcal{H}(L,t) = \mathcal{H}_{out}(t) \end{cases} \tag{7.21}$$

and output values are

$$\begin{cases} \mathcal{H}(0,t) = \mathcal{H}_{in}(t) \\ \mathbb{Q}(L,t) = \mathbb{Q}_{out}(t) \end{cases} \tag{7.22}$$

The vectors $u(t)$ and $y(t)$ contain the system inputs and outputs respectively. In the suggested model, the inputs and outputs are $x = \left(\mathbb{Q}_1 \mathbb{Q}_{2\mathrm{b}} \mathbb{Q}_{\mathrm{n}} \mathcal{H}_0 \mathcal{H}_1 \mathcal{H}_{2\mathrm{b}} \lambda \; \mathrm{z} \; \mathring{A}_b\right)^T$, $u := [\mathbb{Q}_{in} \mathcal{H}_{out}]^T = (\mathbb{Q}_0 \mathcal{H}_{\mathrm{n}})^T$ and $y := (\mathbb{Q}_n \mathcal{H}_0)^T$, respectively. We have,

$$\begin{aligned}
\dot{\mathbb{Q}}_1 &= -\mathring{A} g \frac{\mathcal{H}_1 - \mathcal{H}_0}{z} - \frac{\daleth}{2\eth} \frac{\mathbb{Q}_1^2}{\mathring{A}} \\
\dot{\mathbb{Q}}_{2b} &= -\mathring{A}_b g \frac{(\mathcal{H}_{2b} - \mathcal{H}_{1b})}{\lambda} - \frac{\daleth_b}{2\eth} \frac{\mathbb{Q}_{2b}^2}{\mathring{A}_b} \\
\dot{\mathbb{Q}}_n &= -\mathring{A} g \frac{(\mathcal{H}_n - \mathcal{H}_2)}{(L - z - \lambda)} - \frac{\daleth}{2\eth} \frac{\mathbb{Q}_n^2}{\mathring{A}} \\
\dot{\mathcal{H}}_0 &= -\frac{a^2}{g\mathring{A}} \frac{\mathbb{Q}_1 - \mathbb{Q}_0}{z} \\
\dot{\mathcal{H}}_1 &= -\frac{a^2}{g\mathring{A}} \frac{\mathbb{Q}_1 - \mathbb{Q}_0}{z} \\
\dot{\mathcal{H}}_{2b} &= -\frac{a^2}{g\mathring{A}_b} \frac{(\mathbb{Q}_{2b} - \mathbb{Q}_{1b})}{\lambda} \\
\dot{\lambda} &= 0 \\
\dot{z} &= 0 \\
\dot{\mathring{A}}_b &= 0
\end{aligned} \tag{7.23}$$

where $\mathcal{H}_{1b}$ and $\mathcal{H}_2$ are calculated using Bernoulli's and continuity equations as below,

$$\begin{aligned}
\mathcal{H}_{1b} &= \mathcal{H}_1 + \xi \frac{\mathring{A}_b^2 \mathbb{Q}_1^2 - \mathring{A}^2 \mathbb{Q}_1^2}{2g A^2 A_b^2} \\
\mathcal{H}_2 &= \mathcal{H}_{2b} + \xi \frac{\mathring{A}^2 \mathbb{Q}_{2b}^2 - \mathring{A}_b^2 \mathbb{Q}_{2b}^2}{2g \mathring{A}^2 \mathring{A}_b^2}
\end{aligned} \tag{7.24}$$

Case 3: The initial and boundary conditions that can be controlled and measured are the flow rate and pressure head in the beginning the pipeline, see Fig. 7.3C. The input conditions are stated as

$$\begin{cases} \mathcal{H}(0,t) = \mathcal{H}_{in}(t) \\ \mathbb{Q}(0,t) = \mathbb{Q}_{in}(t) \end{cases} \tag{7.25}$$

and output values are

$$\begin{cases} \mathbb{Q}(L,t) = \mathbb{Q}_{out}(t) \\ \mathcal{H}(L,t) = \mathcal{H}_{out}(t) \end{cases} \tag{7.26}$$

The vectors $u(t)$ and $y(t)$ contain the system inputs and outputs respectively. In the suggested model, the inputs and outputs are $x = \left(\mathbb{Q}_1 \mathbb{Q}_{2\text{b}} \mathbb{Q}_n \mathcal{H}_1 \mathcal{H}_{2\text{b}} \mathcal{H}_n \lambda \text{ z } \mathring{A}_b\right)^T$, $u := [\mathbb{Q}_{in}\mathcal{H}_{in}]^T = (\mathbb{Q}_0\mathcal{H}_0)^T$ and $y := (\mathbb{Q}_\text{n}\mathcal{H}_n)^T$, respectively. We have,

$$\begin{aligned}
\dot{\mathbb{Q}}_1 &= -\mathring{A}g\frac{\mathcal{H}_1 - \mathcal{H}_0}{z} - \frac{\Finv}{2\eth}\frac{\mathbb{Q}_1^2}{\mathring{A}} \\
\dot{\mathbb{Q}}_{2b} &= -\mathring{A}_b g\frac{(\mathcal{H}_{2b} - \mathcal{H}_{1b})}{\lambda} - \frac{\Finv_b}{2\eth}\frac{\mathbb{Q}_{2b}^2}{\mathring{A}_b} \\
\dot{\mathbb{Q}}_n &= -\mathring{A}g\frac{(\mathcal{H}_n - \mathcal{H}_2)}{(L - z - \lambda)} - \frac{\Finv}{2\eth}\frac{\mathbb{Q}_n^2}{\mathring{A}} \\
\dot{\mathcal{H}}_1 &= -\frac{a^2}{g\mathring{A}}\frac{\mathbb{Q}_1 - \mathbb{Q}_0}{z} \\
\dot{\mathcal{H}}_{2b} &= -\frac{a^2}{g\mathring{A}_b}\frac{(\mathbb{Q}_{2b} - \mathbb{Q}_{1b})}{\lambda} \\
\dot{\mathcal{H}}_n &= -\frac{a^2}{g\mathring{A}}\frac{(\mathbb{Q}_n - \mathbb{Q}_2)}{(L - z - \lambda)} \\
\dot{\lambda} &= 0 \\
\dot{z} &= 0 \\
\dot{\mathring{A}}_b &= 0
\end{aligned} \tag{7.27}$$

where $\mathcal{H}_{1b}$ and $\mathcal{H}_2$ are calculated using Bernoulli's and continuity equations as below,

$$\begin{aligned}
\mathcal{H}_{1b} &= \mathcal{H}_1 + \xi\frac{\mathring{A}_b^2\mathbb{Q}_1^2 - \mathring{A}^2\mathbb{Q}_1^2}{2g\mathring{A}^2\mathring{A}_b^2} \\
\mathcal{H}_2 &= \mathcal{H}_{2b} + \xi\frac{\mathring{A}^2\mathbb{Q}_{2b}^2 - \mathring{A}_b^2\mathbb{Q}_{2b}^2}{2g\mathring{A}^2\mathring{A}_b^2}
\end{aligned} \tag{7.28}$$

Case 4: The initial and boundary conditions that can be controlled and measured are the flow rate in the beginning and pressure head in end of the pipeline, see

Fig. 7.3D. The input conditions are stated as

$$\begin{cases} \mathcal{H}(\mathrm{L}, t) = \mathcal{H}_{out}(t) \\ \mathbb{Q}(0, t) = \mathbb{Q}_{in}(t) \end{cases} \tag{7.29}$$

and output values are

$$\begin{cases} \mathbb{Q}(L, t) = \mathbb{Q}_{out}(t) \\ \mathcal{H}(0, t) = \mathcal{H}_{in}(t) \end{cases} \tag{7.30}$$

The vectors $u(t)$ and $y(t)$ contain the system inputs and outputs respectively. In the suggested model, the inputs and outputs are $x = \left(\mathbb{Q}_1 \mathbb{Q}_{2b} \mathbb{Q}_n \mathcal{H}_0 \mathcal{H}_1 \mathcal{H}_{2b} \lambda \; \mathrm{z} \; \mathring{A}_b\right)^T$, $u := [\mathbb{Q}_{in} \mathcal{H}_{out}]^T = (\mathbb{Q}_0 \mathcal{H}_n)^T$ and $y := (\mathbb{Q}_n \mathcal{H}_0)^T$, respectively. We have,

$$\begin{aligned}
\dot{\mathbb{Q}}_1 &= -\mathring{A} g \frac{\mathcal{H}_1 - \mathcal{H}_0}{z} - \frac{\Finv}{2\eth} \frac{\mathbb{Q}_1^2}{\mathring{A}} \\
\dot{\mathbb{Q}}_{2b} &= -\mathring{A}_b g \frac{(\mathcal{H}_{2b} - \mathcal{H}_{1b})}{\lambda} - \frac{\Finv_b}{2\eth} \frac{\mathbb{Q}_{2b}^2}{\mathring{A}_b} \\
\dot{\mathbb{Q}}_n &= -\mathring{A} g \frac{(\mathcal{H}_n - \mathcal{H}_2)}{(L - z - \lambda)} - \frac{\Finv}{2\eth} \frac{\mathbb{Q}_n^2}{\mathring{A}} \\
\dot{\mathcal{H}}_0 &= -\frac{a^2}{g\mathring{A}} \frac{\mathbb{Q}_1 - \mathbb{Q}_0}{z} \\
\dot{\mathcal{H}}_1 &= -\frac{a^2}{g\mathring{A}} \frac{\mathbb{Q}_1 - \mathbb{Q}_0}{z} \\
\dot{\mathcal{H}}_{2b} &= -\frac{a^2}{g\mathring{A}_b} \frac{(\mathbb{Q}_{2b} - \mathbb{Q}_{1b})}{\lambda} \\
\dot{\lambda} &= 0 \\
\dot{z} &= 0 \\
\dot{\mathring{A}}_b &= 0
\end{aligned} \tag{7.31}$$

where $\mathcal{H}_{1b}$ and $\mathcal{H}_2$ are calculated using Bernoulli's and continuity equations as below,

$$\begin{aligned}
\mathcal{H}_{1b} &= \mathcal{H}_1 + \xi \frac{\mathring{A}_b^2 \mathbb{Q}_1^2 - \mathring{A}^2 \mathbb{Q}_1^2}{2g A^2 A_b^2} \\
\mathcal{H}_2 &= \mathcal{H}_{2b} + \xi \frac{\mathring{A}^2 \mathbb{Q}_{2b}^2 - \mathring{A}_b^2 \mathbb{Q}_{2b}^2}{2g \mathring{A}^2 \mathring{A}_b^2}
\end{aligned} \tag{7.32}$$

Case 5: The initial and boundary conditions that can be controlled and measured are the flow rate in the beginning and end of the pipeline, see Fig. 7.3E. The input conditions are stated as

$$\begin{cases} \mathbb{Q}(0,t) = \mathbb{Q}_{in}(t) \\ \mathbb{Q}(L,t) = \mathbb{Q}_{out}(t) \end{cases} \tag{7.33}$$

and output values are

$$\begin{cases} \mathcal{H}(0,t) = \mathcal{H}_{in}(t) \\ \mathcal{H}(L,t) = \mathcal{H}_{out}(t) \end{cases} \tag{7.34}$$

The vectors $u(t)$ and $y(t)$ contain the system inputs and outputs respectively. In the suggested model, the inputs and outputs are $x = \left(\mathbb{Q}_1 \mathbb{Q}_{2\text{b}} \mathcal{H}_0 \mathcal{H}_1 \mathcal{H}_{2\text{b}} \mathcal{H}_n \lambda \text{ z } \mathring{A}_b\right)^T$, $u := [\mathbb{Q}_{in} \mathbb{Q}_{out}]^T = (\mathbb{Q}_0 \mathbb{Q}_n)^T$ and $y := (\mathcal{H}_0 \mathcal{H}_n)^T$, respectively.

$$\begin{aligned}
\dot{\mathbb{Q}}_1 &= -\mathring{A} g \frac{\mathcal{H}_1 - \mathcal{H}_0}{z} - \frac{\daleth}{2\eth} \frac{\mathbb{Q}_1^2}{\mathring{A}} \\
\dot{\mathbb{Q}}_{2b} &= -\mathring{A}_b g \frac{(\mathcal{H}_{2b} - \mathcal{H}_{1b})}{\lambda} - \frac{\daleth_b}{2\eth} \frac{\mathbb{Q}_{2b}^2}{\mathring{A}_b} \\
\dot{\mathcal{H}}_0 &= -\frac{a^2}{g\mathring{A}} \frac{\mathbb{Q}_1 - \mathbb{Q}_0}{z} \\
\dot{\mathcal{H}}_1 &= -\frac{a^2}{g\mathring{A}} \frac{\mathbb{Q}_1 - \mathbb{Q}_0}{z} \\
\dot{\mathcal{H}}_{2b} &= -\frac{a^2}{g\mathring{A}_b} \frac{(\mathbb{Q}_{2b} - \mathbb{Q}_{1b})}{\lambda} \\
\dot{\mathcal{H}}_n &= -\frac{a^2}{g\mathring{A}} \frac{(\mathbb{Q}_n - \mathbb{Q}_2)}{(L - z - \lambda)} \\
\dot{\lambda} &= 0 \\
\dot{z} &= 0 \\
\dot{\mathring{A}}_b &= 0
\end{aligned} \tag{7.35}$$

where H_{1b} and H_2 are calculated using Bernoulli's and continuity equations as below,

$$\begin{aligned}
\mathcal{H}_{1b} &= \mathcal{H}_1 + \xi \frac{\mathring{A}_b^2 \mathbb{Q}_1^2 - \mathring{A}^2 \mathbb{Q}_1^2}{2g A^2 A_b^2} \\
\mathcal{H}_2 &= \mathcal{H}_{2b} + \xi \frac{\mathring{A}^2 \mathbb{Q}_{2b}^2 - \mathring{A}_b^2 \mathbb{Q}_{2b}^2}{2g \mathring{A}^2 \mathring{A}_b^2}
\end{aligned} \tag{7.36}$$

Case 6: The boundary conditions that can be controlled and measured are the pressure head and the flow rate in the beginning and end of the pipeline, respectively, see Fig. 7.3F. The input conditions are stated as

$$\begin{cases} \mathbb{Q}(L,t) = \mathbb{Q}_{out}(t) \\ \mathcal{H}(\mathrm{L},t) = \mathcal{H}_{out}(t) \end{cases} \tag{7.37}$$

and output values are

$$\begin{cases} \mathbb{Q}(0,t) = \mathbb{Q}_{in}(t) \\ \mathcal{H}(L,t) = \mathcal{H}_{in}(t) \end{cases} \tag{7.38}$$

The vectors $u(t)$ and $y(t)$ contain the system inputs and outputs respectively. In the suggested model, the inputs and outputs are $x = \left(\mathbb{Q}_1\mathbb{Q}_{2\mathrm{b}}\mathbb{Q}_n\mathcal{H}_0\mathcal{H}_1\mathcal{H}_{2\mathrm{b}}\lambda\ \mathrm{z}\ \mathring{A}_b\right)^T$, $u := [\mathbb{Q}_{out}\mathcal{H}_{out}]^T = (\mathbb{Q}_{\mathrm{n}}\mathcal{H}_n)^T$ and $y := (\mathbb{Q}_0\mathcal{H}_0)^T$, respectively.

$$\begin{aligned}
\dot{\mathbb{Q}}_0 &= -\mathring{A}g\frac{\mathcal{H}_1 - \mathcal{H}_0}{z} - \frac{\daleth}{2\eth}\frac{\mathbb{Q}_0^2}{\mathring{A}} \\
\dot{\mathbb{Q}}_1 &= -\mathring{A}g\frac{\mathcal{H}_1 - \mathcal{H}_0}{z} - \frac{\daleth}{2\eth}\frac{\mathbb{Q}_1^2}{\mathring{A}} \\
\dot{\mathbb{Q}}_{2b} &= -\mathring{A}_b g\frac{(\mathcal{H}_{2b} - \mathcal{H}_{1b})}{\lambda} - \frac{\daleth_b}{2\eth}\frac{\mathbb{Q}_{2b}^2}{\mathring{A}_b} \\
\dot{\mathcal{H}}_0 &= -\frac{a^2}{g\mathring{A}}\frac{\mathbb{Q}_1 - \mathbb{Q}_0}{z} \\
\dot{\mathcal{H}}_1 &= -\frac{a^2}{g\mathring{A}}\frac{\mathbb{Q}_1 - \mathbb{Q}_0}{z} \\
\dot{\mathcal{H}}_{2b} &= -\frac{a^2}{g\mathring{A}_b}\frac{(\mathbb{Q}_{2b} - \mathbb{Q}_{1b})}{\lambda} \\
\dot{\lambda} &= 0 \\
\dot{z} &= 0 \\
\dot{\mathring{A}}_b &= 0
\end{aligned} \tag{7.39}$$

where $\mathcal{H}_{1b}$ and $\mathcal{H}_2$ are calculated using Bernoulli's and continuity equations as below,

$$\begin{aligned}
\mathcal{H}_{1b} &= \mathcal{H}_1 + \xi\frac{\mathring{A}_b^2\mathbb{Q}_1^2 - \mathring{A}^2\mathbb{Q}_1^2}{2gA^2A_b^2} \\
\mathcal{H}_2 &= \mathcal{H}_{2b} + \xi\frac{\mathring{A}^2\mathbb{Q}_{2b}^2 - \mathring{A}_b^2\mathbb{Q}_{2b}^2}{2g\mathring{A}^2\mathring{A}_b^2}
\end{aligned} \tag{7.40}$$

A mathematical expression for a leak can be deduced from Bernoulli's equation, which relates the pressure difference between the pipeline inside and outside.

7.3 Observer Design by Using the Extended Kalman Filter

The idea considered in this chapter for the aim of blockage identification problem is the blockage parameter approximation z, λ as well as $\mathring{A}_b$, utilizing an Extended Kalman Filter as a state observer. For this aim, the model proposed in (7.15) and (7.16) can be developed with the dynamics of these parameters, also a nonlinear observer can be made for the developed system [63]. This needs available measurements and input variables, be determined. For different cases it has been assumed to measure just pressures head, just flow rates or combination of both at the beginning or end of the pipeline. For case 1, flow rates in the beginning and end of the pipeline are considered as outputs which are directly measured as follows,

$$y = [\mathbb{Q}_0 \mathbb{Q}_n]^T \tag{7.41}$$

and the pressure heads in the beginning and end of pipeline considered as known inputs are

$$u = [\mathcal{H}_0 \mathcal{H}_n]^T \tag{7.42}$$

Finally, considering extensions

$$\dot{z} = 0; \dot{\lambda} = 0; \dot{\mathring{A}}_b = 0. \tag{7.43}$$

The derivative of blockage parameters z, λ and $\mathring{A}_b$ can be taken as zero since their variation is small. Therefore (7.15) and (7.16) can be extended as below,

$$\begin{bmatrix} \dot{\mathbb{Q}}_0 \\ \dot{\mathbb{Q}}_1 \\ \dot{\mathbb{Q}}_{2b} \\ \dot{\mathbb{Q}}_n \\ \dot{\mathcal{H}}_1 \\ \dot{\mathcal{H}}_{2b} \\ \dot{\lambda} \\ \dot{z} \\ \dot{\mathring{A}}_b \end{bmatrix} = \begin{bmatrix} -\mathring{A} g \frac{\mathcal{H}_1 - \mathcal{H}_0}{z} - \frac{\Finv}{2\eth} \frac{\mathbb{Q}_0^2}{\mathring{A}} \\ -\mathring{A} g \frac{\mathcal{H}_1 - \mathcal{H}_0}{z} - \frac{\Finv}{2\eth} \frac{\mathbb{Q}_1^2}{\mathring{A}} \\ -\mathring{A}_b g \frac{(\mathcal{H}_{2b} - \mathcal{H}_{1b})}{\lambda} - \frac{\Finv_b}{2\eth} \frac{\mathbb{Q}_{2b}^2}{\mathring{A}_b} \\ -\mathring{A} g \frac{(\mathcal{H}_n - \mathcal{H}_2)}{(L - z - \lambda)} - \frac{\Finv}{2\eth} \frac{\mathbb{Q}_n^2}{\mathring{A}} \\ -\frac{a^2}{g\mathring{A}} \frac{\mathbb{Q}_1 - \mathbb{Q}_0}{z} \\ -\frac{a^2}{g\mathring{A}_b} \frac{(\mathbb{Q}_{2b} - \mathbb{Q}_{1b})}{\lambda} \\ 0 \\ 0 \\ 0 \end{bmatrix} \tag{7.44}$$

Hence, the model (7.44) can be written in compact form as:

$$\dot{x} = \Psi(x, u) \tag{7.45}$$

where $x = \left(\mathbb{Q}_1 \mathbb{Q}_{2b} \mathbb{Q}_n \mathcal{H}_0 \mathcal{H}_1 \mathcal{H}_{2b} \lambda \; \mathrm{z} \; \mathring{A}_b\right)^T$ and $\Psi(x, u)$ is taken as a nonlinear function.

Now, to estimate the leak parameters z, λ and $\mathring{A}_0$, a discrete-time Extended Kalman Filter is designed for the nonlinear model described in (7.44). To do that, this model is discretized by using Heun's method. In this method, the solution for the initial value problem is given by [64],

$$\dot{x} = \Psi(x(t), u(t)), x(t_0) = x_0 \tag{7.46}$$

Therefore,

$$x^{i+1} = x^i + \frac{\Delta t}{2}\left[\phi\left(x^i, u^i\right) + \Psi\left(x^i + \Delta t \Psi\left(x^i, u^i\right), u^{i+1}\right)\right] \tag{7.47}$$

where Δt is the time step and i is the index of discrete-time. Finally, the ordinary differential equations in (7.44) are transformed by using (7.47), into the following difference equations:

$$\begin{aligned}
\mathbb{Q}_0^{i+1} &= \mathbb{Q}_0^i + \frac{\Delta t}{2}\left(-\mathring{A}g\frac{H_1^i - H_0^i}{z^i} - \frac{\text{Ⅎ}}{2\eth}\frac{\mathbb{Q}_0^i}{\mathring{A}} - \mathring{A}g\frac{\hat{\mathcal{H}}_1^{i+1} - \mathcal{H}_0^{i+1}}{z^{i+1}}\right) - \frac{\text{Ⅎ}}{2\eth}\frac{\hat{\mathbb{Q}}_0^{i+1}}{\mathring{A}} \\
\mathbb{Q}_1^{i+1} &= \mathbb{Q}_1^i + \frac{\Delta t}{2}\left(-\mathring{A}g\frac{\mathcal{H}_1^i - \mathcal{H}_0^i}{z^i} - \frac{\text{Ⅎ}}{2\eth}\frac{\mathbb{Q}_1^i}{\mathring{A}} - \mathring{A}g\frac{\hat{\mathcal{H}}_1^{i+1} - \hat{\mathcal{H}}_0^{i+1}}{z^{i+1}}\right) - \frac{\text{Ⅎ}}{2\eth}\frac{\hat{\mathbb{Q}}_1^{i+1}}{\mathring{A}} \\
\mathbb{Q}_{2b}^{i+1} &= \mathbb{Q}_{2b}^i + \frac{\Delta t}{2}\left(-\mathring{A}_b g\frac{\hat{\mathcal{H}}_{2b}^i - \hat{\mathcal{H}}_{1b}^i}{\lambda^i} - \frac{\text{Ⅎ}_b}{2\eth}\frac{\mathbb{Q}_{2b}^i}{\mathring{A}_b} - \mathring{A}g\frac{\hat{\mathcal{H}}_{2b}^{i+1} - \hat{\mathcal{H}}_{1b}^{i+1}}{\lambda^{i+1}}\right) - \frac{\text{Ⅎ}}{2\eth}\frac{\hat{\mathbb{Q}}_{2b}^{i+1}}{\mathring{A}} \\
\mathbb{Q}_n^{i+1} &= \mathbb{Q}_n^i + \frac{\Delta t}{2}\left(-\mathring{A}g\frac{\hat{\mathcal{H}}_n^i - \hat{\mathcal{H}}_2^i}{\left(L - z^i - \lambda^i\right)} - \frac{\text{Ⅎ}}{2\eth}\frac{\mathbb{Q}_n^i}{\mathring{A}} - \mathring{A}g\frac{\hat{\mathcal{H}}_n^{i+1} - \hat{\mathcal{H}}_2^{i+1}}{\left(L - z^{i+1} - \lambda^{i+1}\right)}\right) \\
&\quad - \frac{\text{Ⅎ}}{2\eth}\frac{\hat{\mathbb{Q}}_n^{i+1}}{\mathring{A}} \\
\mathcal{H}_1^{i+1} &= \mathcal{H}_n^i + \frac{\Delta t}{2}\left(-\frac{a^2}{gA}\frac{\mathbb{Q}_1^i - \mathbb{Q}_0^i}{z^i} - \frac{a^2}{gA}\frac{\hat{\mathbb{Q}}_1^{i+1} - \hat{\mathbb{Q}}_0^{i+1}}{z^{i+1}}\right) \\
\mathcal{H}_{2b}^{i+1} &= \mathcal{H}_{2b}^i + \frac{\Delta t}{2}\left(-\frac{a^2}{gA_b}\frac{\mathbb{Q}_1^i - \mathbb{Q}_0^i}{\lambda^i} - \frac{a^2}{g\mathring{A}}\frac{\hat{\mathbb{Q}}_1^{i+1} - \hat{\mathbb{Q}}_0^{i+1}}{\lambda^{i+1}}\right) \\
\lambda^{i+1} &= \lambda^i \\
z^{i+1} &= z^i \\
\mathring{A}_b^{i+1} &= \mathring{A}_b^i
\end{aligned} \tag{7.48}$$

in compact form:

$$x^{i+1} = \Psi\left(x^i, u^{i+1}, u^i\right);\ y^i = Hx^i \tag{7.49}$$

where

$$\begin{aligned} x^i &= \left[\mathbb{Q}_0^i \mathbb{Q}_1^i \mathbb{Q}_{2b}^i \mathbb{Q}_n^i \mathcal{H}_n^i \mathcal{H}_{2b}^i \lambda^i z^i A_b^i\right]^T; \\ u^i &= \left[\mathcal{H}_0^i \mathcal{H}_n^i\right]^T \\ H &= \begin{bmatrix} 1\,0\,0\;0\,0\,0\;0\,0\,0 \\ 0\,0\,0\;1\,0\,0\;0\,0\,0 \end{bmatrix} \end{aligned} \tag{7.50}$$

7.3.1 Observer Scheme

A discrete-time Extended Kalman Filter as state observer for the system (7.44) can be chosen as [65],

$$\hat{x}^i = \hat{x}^{\tilde{i}} + \kappa^i\left(y^i - H\hat{x}^{\tilde{i}}\right) \tag{7.51}$$

where $\hat{x}^{\tilde{i}}$ is the priori estimate of x^i:

$$\hat{x}^{\tilde{i}} = \Psi\left(\hat{x}^{i-1}, u^{i-1}\right) \tag{7.52}$$

κ^i is the Kalman gain:

$$\kappa^i = P^{\tilde{i}} H^T\left(HP^{\tilde{i}}H^T + R\right)^{-1} \tag{7.53}$$

$P^{\tilde{i}}$ is the priori covariance matrix:

$$P^{\tilde{i}} = J^i P^{i-1}\left(J^i\right)^{-1} \tag{7.54}$$

P^i is the posteriori covariance matrix:

$$P^i = \left(I - K^i H\right)P^{\tilde{i}} \tag{7.55}$$

J^i is the Jacobian matrix:

$$J^i = \frac{\partial\phi(x, u)}{\partial x}|_{x=\hat{x}^i} \tag{7.56}$$

Finally, R and D are known as the covariance matrices of measure and process noises, respectively. Notice that:

$$\hat{x}^i = \left(\hat{\mathbb{Q}}_1^i \hat{\mathbb{Q}}_{2b}^i \hat{\mathbb{Q}}_n^i \hat{\mathcal{H}}_0^i \hat{\mathcal{H}}_1^i \hat{\mathcal{H}}_{2b}^i \hat{\lambda}^i \hat{z}^i \hat{\text{Å}}_b^i\right) \tag{7.57}$$

with $P^{\tilde{0}} = \left(P^{\tilde{0}}\right)^T > 0$, $R = R^T > 0$ and $D = D^T > 0$.

7.4 Simulation Results

In this section, simulations results are reported for one case.

The model has been realized on the pipeline with the following physical parameters:

The length of the pipeline is $L = 1.2 \times 10^3$ m, the diameter of the pipe is $\eth = 0.56\,\text{m}$, the cross-section is $\text{Å} = 0.246\,\text{m}^2$, density is $\rho = 1000\,\text{kg/m}^3$, gravity is $g = 9.81\,\text{m/s}^2$, the friction factor of the pipe is $\daleth = 0.006$, the friction factor of the blockage part is $\daleth_b = 0.016$ and the wave speed is $a = 1250\,\text{m/s}$ (Fig. 7.4).

Control variables for Case 1 model are the pressures head at the beginning and end of the pipe ($\mathcal{H}_0$ and $\mathcal{H}_n$) and can be directly measured. Furthermore, the flow rates (Q_0 and Q_n) are outputs of the system.

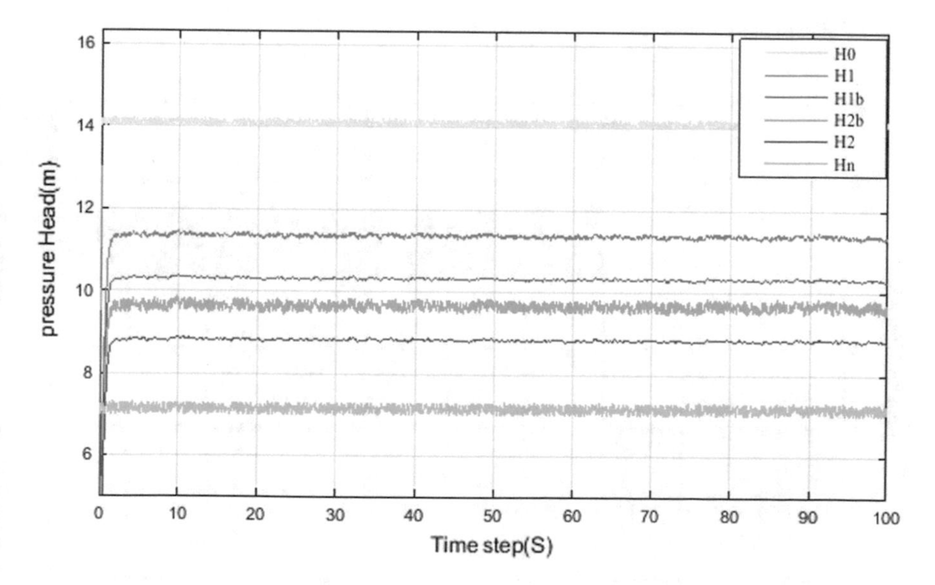

Fig. 7.4 Pressure head in pipeline

The simulation has the same structure as the designed system demonstrated in (7.15) and (7.16), the simulate pressure head at the inlet ($\mathcal{H}(in) = \mathcal{H}_0 = 16\,\text{m}$) and outlet ($\mathcal{H}(out) = \mathcal{H}_n = 5.5\,\text{m}$) of the pipe.

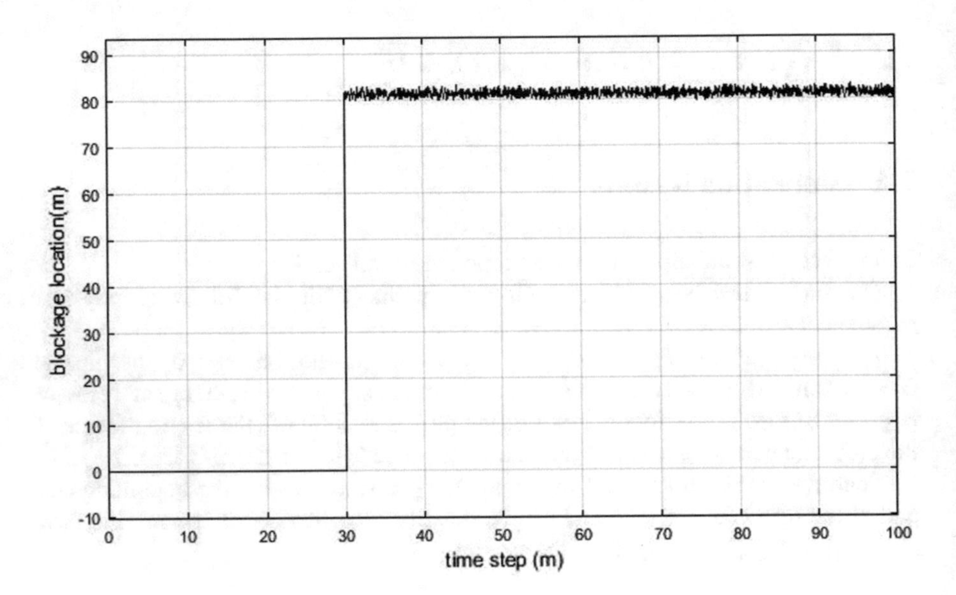

Fig. 7.5 Position of blockage

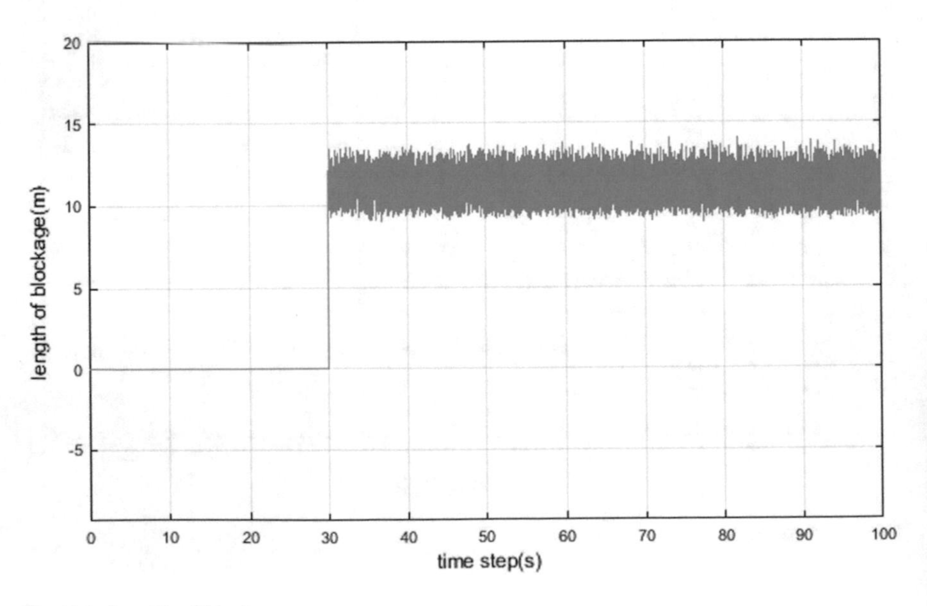

Fig. 7.6 Length of blockage

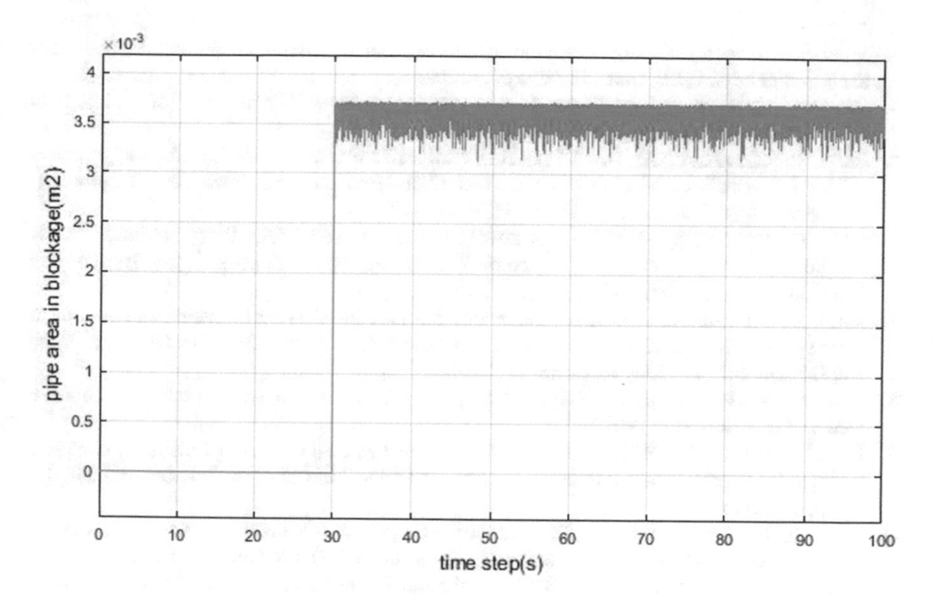

Fig. 7.7 Pipe cross section area in blockage

Figures 7.5, 7.6 and 7.7 estimate the position, length and the area of blockage in the pipeline.

7.5 Conclusion

The objective of this chapter is to analyze and model the blockage in the pipeline system. The system is stated by hyperbolic partial differential equations. In this chapter, the finite difference technique is implemented since it is a simple approach to design a more convenient model for observing and controlling the structure of the nonlinear system. This technique partitions the whole pipeline into N number of sections. Flow and pressure head estimations have been well estimated in the presence of a blockage. Some theorems are presented in order to detect blockage in the system. Future work is to study the case of multi blockage estimation of the pipeline system.

References

1. Yu, W., Jafari, R.: Modeling and Control of Uncertain Nonlinear Systems with Fuzzy Equations and Z-Number. Wiley (2019)
2. Razvarz, S., Jafari, R.: ICA and ANN modeling for photocatalytic removal of pollution in wastewater. Math. Comput. Appl. **22**(3), 38 (2017)

3. Jafari, R., Razvarz, S., Gegov, A.: Neural network approach to solving fuzzy nonlinear equations using Z-numbers. IEEE Trans. Fuzzy Syst. (2019)
4. Razvarz, S., Jafari, R.: Intelligent techniques for photocatalytic removal of pollution in wastewater. J. Electr. Eng. **5**(1), 321–328 (2017)
5. Jafari, R., Razvarz, S., Gegov, A., Paul, S., Keshtkar, S.: Fuzzy Sumudu transform approach to solving fuzzy differential equations with Z-numbers. In: Advanced Fuzzy Logic Approaches in Engineering Science, pp. 18–48. IGI Global (2019)
6. Jafari, R., Razvarz, S., Gegov, A.: A novel technique to solve fully fuzzy nonlinear matrix equations. In: International Conference on Theory and Applications of Fuzzy Systems and Soft Computing, pp. 886–892. Springer (2018)
7. Jafari, R., Razvarz, S., Gegov, A.: Fuzzy differential equations for modeling and control of fuzzy systems. In: International Conference on Theory and Applications of Fuzzy Systems and Soft Computing, pp. 732–740. Springer (2018)
8. Jafari, R., Yu, W., Razvarz, S., Gegov, A.: Numerical methods for solving fuzzy equations: a survey. Fuzzy Sets Syst. (2019)
9. Jafari, R., Razvarz, S., Gegov, A.: A new computational method for solving fully fuzzy nonlinear systems. In: International Conference on Computational Collective Intelligence, pp. 503–512. Springer (2018)
10. Jafari, R., Razvarz, S., Gegov, A., Paul, S.: Modeling and control of uncertain nonlinear systems. In: 2018 International Conference on Intelligent Systems (IS), pp. 168–173. IEEE (2018)
11. Jafari, R., Razvarz, S., Gegov, A.: A novel technique for solving fully fuzzy nonlinear systems based on neural networks. Vietnam J. Comput. Sci. **7**(1), 93–107 (2020)
12. Razvarz, S., Hernández-Rodríguez, F., Jafari, R., Gegov, A.: Foundation of Z-numbers and engineering applications. In: Latin American Symposium on Industrial and Robotic Systems, pp. 15–24. Springer (2019)
13. Jafari, R., Contreras, M.A., Yu, W., Gegov, A.: Applications of fuzzy logic, artificial neural network and neuro-fuzzy in industrial engineering. In: Latin American Symposium on Industrial and Robotic Systems, pp. 9–14. Springer (2019)
14. Jafari, R., Razvarz, S., Gegov, A., Yu, W.: Fuzzy control of uncertain nonlinear systems with numerical techniques: a survey. In: UK Workshop on Computational Intelligence, pp. 3–14. Springer (2019)
15. Jafari, R., Razvarz, S., Yu, W., Gegov, A., Goodwin, M., Adda, M.: Genetic algorithm modeling for photocatalytic elimination of impurity in wastewater. In: Proceedings of SAI Intelligent Systems Conference, pp. 228–236. Springer (2019)
16. Tatchum, M., Gegov, A., Jafari, R., Razvarz, S.: Parallel distributed compensation for voltage controlled active magnetic bearing system using integral fuzzy model. In: 2018 International Conference on Intelligent Systems (IS), pp. 190–198. IEEE (2018)
17. Razvarz, S., Jafari, R., Gegov, A.: Solving partial differential equations with Bernstein neural networks. In: UK Workshop on Computational Intelligence, pp. 57–70. Springer (2018)
18. Jafarian, A., Jafari, R.: New iterative approach for solving fully fuzzy polynomials. Int. J. Fuzzy Math. Syst. **3**(2), 75–83
19. Jafarian, A., Jafari, R.: New method for solving fuzzy polynomials. Adv. Fuzzy Math. **8**(1), 25–33 (2013)
20. Jafarian, A., Jafari, R.: An iterative method for solving fuzzy polynomials by fuzzy neural networks (2012)
21. Jafarian, A., Jafari, R.: Simulation and evaluation of fuzzy polynomials by feed-back neural networks (2012)
22. Jafari, R., Yu, W.: Fuzzy control for uncertainty nonlinear systems with dual fuzzy equations. J. Intell. Fuzzy Syst. **29**(3), 1229–1240 (2015)
23. Jafari, R., Yu, W.: Fuzzy modeling for uncertainty nonlinear systems with fuzzy equations. Math. Probl. Eng. (2017)
24. Verde, C.: Multi-leak detection and isolation in fluid pipelines. Control Eng. Pract. **9**(6), 673–682 (2001)

25. Kowalczuk, Z., Gunawickrama, K.: Leak detection and isolation for transmission pipelines via nonlinear state estimation. IFAC Proc. Vol. **33**(11), 921–926 (2000)
26. Verde, C., Visairo, N.: Multi-leak isolation in a pipeline by unsteady state test. In: 44th IEEE Conference on Decision and Control and European Control Conference ECC, pp. 1711–1721 (2005)
27. Ghazali, M., Beck, S., Shucksmith, J., Boxall, J., Staszewski, W.: Comparative study of instantaneous frequency based methods for leak detection in pipeline networks. Mech. Syst. Signal Process. **29**, 187–200 (2012)
28. Mpesha, W., Gassman, S.L., Chaudhry, M.H.: Leak detection in pipes by frequency response method. J. Hydraul. Eng. **127**(2), 134–147 (2001). https://doi.org/10.1061/(ASCE)0733-9429(2001)127:2(134)
29. Covas, D., Ramos, H., De Almeida, A.B.: Standing wave difference method for leak detection in pipeline systems. J. Hydraul. Eng. **131**(12), 1106–1116 (2005)
30. Billmann, L., Isermann, R.: Leak detection methods for pipelines. Automatica **23**(3), 381–385 (1987)
31. Verde, C., Torres, L., González, O.: Decentralized scheme for leaks' location in a branched pipeline. J. Loss Prev. Process Ind. **43**, 18–28 (2016)
32. Verde, C., Visairo, N.: Identificability of multi-leaks in a pipeline. In: Proceedings of the 2004 American Control Conference, pp. 4378–4383. IEEE (2004)
33. Verde, C., Visairo, N., Gentil, S.: Two leaks isolation in a pipeline by transient response. Adv. Water Resour. **30**(8), 1711–1721 (2007)
34. Duan, H.F., Lee, P.J., Ghidaoui, M.S., Tung, Y.K.: Extended blockage detection in pipelines by using the system frequency response analysis. J. Water Resour. Plan. Manag. **138**(1), 55–62 (2012)
35. Mushiri, T., Ndlovu, S., Mbohwa, C.: Design of a mechanical cleaning device PIG (pipeline intervention gadget) connecting two transfer lines in Zimbabwe (2016)
36. Kanniga, E., Malik, M.: Survey on smart pig for inspection robot. Int. J. Psychosoc. Rehabil. **23**(3) (2019)
37. Moore, D., Moore, G., Magnusson, J., Boase, C.: Pipeline pig. Google Patents (2002)
38. Hunt, H.E., Kopke, U.G.: Pipeline pig and method of pipeline inspection. Google Patents (1995)
39. Davis, G.W.: Method and a horizontal pipeline pig launching mechanism for sequentially launching pipeline pigs. Google Patents (1992)
40. Mirshamsi, M., Rafeeyan, M.: Speed control of pipeline pig using the QFT method. Oil Gas Sci. Technol. Revue d'IFP Energies Nouvelles **67**(4), 693–701 (2012)
41. Adewumi, M.A., Eltohami, E., Ahmed, W.: Pressure transients across constrictions. J. Energy Resour. Technol. **122**(1), 34–41 (2000)
42. Adewumi, M.A., Eltohami, E., Solaja, A.: Possible detection of multiple blockages using transients. J. Energy Resour. Technol. **125**(2), 154–159 (2003)
43. Vítkovsky, J.P., Lee, P.J., Stephens, M.L., Lambert, M.F., Simpson, A.R., Liggett, J.A., Cabrera, E.: Leak and blockage detection in pipelines via an impulse response method. In: Cabrera, E., Jr. (ed.) Pumps, Electromechanical Devices and Systems Applied to Urban Water Management, pp. 423–430 (2003)
44. Wang, X.J., Lambert, M.F., Simpson, A.R.: Detection and location of a partial blockage in a pipeline using damping of fluid transients. J. Water Resour. Plan. Manag. **131**(3), 244–249 (2005)
45. Sattar, A.M., Chaudhry, M.H., Kassem, A.A.: Partial blockage detection in pipelines by frequency response method. J. Hydraul. Eng. **134**(1), 76–89 (2008)
46. Razvarz, S., Vargas-Jarillo, C., Jafari, R.: Pipeline monitoring architecture based on observability and controllability analysis. In: 2019 IEEE International Conference on Mechatronics (ICM), 18–20 Mar 2019, pp. 420–423
47. Jafari, R., Razvarz, S., Vargas-Jarillo, C., Yu, W.: Control of flow rate in pipeline using PID controller. In: 2019 IEEE 16th International Conference on Networking, Sensing and Control (ICNSC), 9–11 May 2019, pp. 293–298

48. Jafari, R., Razvarz, S., Vargas-Jarillo, C., Gegov, A.: Blockage detection in pipeline based on the extended Kalman filter observer. Electronics **9**(1), 91–107 (2020)
49. Razvarz, S., Jafari, R., Vargas-Jarillo, C.: Modelling and analysis of flow rate and pressure head in pipelines. In: 2019 16th International Conference on Electrical Engineering, Computing Science and Automatic Control (CCE), pp. 1–6. IEEE (2019)
50. Jafari, R., Razvarz, S., Vargas-Jarillo, C., Gegov, A.E.: The effect of baffles on heat transfer. In: ICINCO (2), pp. 607–612 (2019)
51. Razvarz, S., Jafari, R., Vargas-Jarillo, C., Gegov, A., Forooshani, M.: Leakage detection in pipeline based on second order extended Kalman filter observer. IFAC-PapersOnLine **52**(29), 116–121 (2019)
52. Razvarz, S., Vargas-Jarillo, C., Jafari, R., Gegov, A.: Flow control of fluid in pipelines using PID controller. IEEE Access **7**, 25673–25680 (2019)
53. Razvarz, S., Chavez, L.F.G., Vargas-Jarillo, C.: Nanotechnology applications in industry and heat transfer. In: Latin American Symposium on Industrial and Robotic Systems, pp. 1–8. Springer (2019)
54. Brown, G.O.: The history of the Darcy-Weisbach equation for pipe flow resistance. In: Environmental and Water Resources History, pp. 34–43 (2003)
55. Allen, R.: Relating the Hazen-Williams and Darcy-Weisbach friction loss equations for pressurized irrigation. Appl. Eng. Agric. **12**(6), 685–693 (1996)
56. Swanee, P., Jain, A.K.: Explicit equations for pipeflow problems. J. Hydraul. Div. **102**(5) (1976)
57. Kiijarvi, J.: Darcy Friction Factor Formulae in Turbulent Pipe Flow. Lunowa Fluid Mechanics Paper 110727, pp. 1–11 (2011)
58. Wylie, E.B., Streeter, V.L., Suo, L.: Fluid Transients in Systems, vol. 1. Prentice Hall, Englewood Cliffs, NJ (1993)
59. Wylie, E.B., Streeter, V.L.: Fluid Transients. MHI (1978)
60. Chaudhry, M.H.: Transient-flow equations. In: Applied Hydraulic Transients, pp. 35–64. Springer (2014)
61. Batchelor, C.K., Batchelor, G.: An Introduction to Fluid Dynamics. Cambridge University Press (2000)
62. Ragheb, M.: Fluid Mechanics, Euler and Bernoulli Equations (2013)
63. Besançon, G.: Nonlinear Observers and Applications, vol. 363. Springer (2007)
64. Golub, G.H., Ortega, J.M.: Scientific Computing and Differential Equations: An Introduction to Numerical Methods. Elsevier (2014)
65. Simon, D.: Optimal State Estimation: Kalman, H Infinity, and Nonlinear Approaches. Wiley (2006)

Chapter 8
Leakage Detection in Pipeline Based on Second Order Extended Kalman Filter Observer

8.1 Introduction

Pipelines are extensively implemented in transferring great amounts of oil and gas products at large distances because of their safety, effectiveness as well as low cost. The protection of pipelines from theft and leakage is one of the primary aims of the oil and gas companies. Piping engineering is one of the fields that by monitoring the pipelines and using the methods such as artificial neural network [1–23], hydrostatic testing, infrared, and laser technology tries to found flaw in pipeline.

Pipelines could suffer from various kinds of defects like corrosion, exhaustion cracks, dent, etc. Therefore, if these defects are not appropriately managed can cause pipeline failures having leak or rupture that may lead to extremely downtime and environment dangers [24–35].

Automatic fault monitoring, as well as the detection in fluid distribution systems, are of great importance in today's world. The main aim of the automatic pipeline monitoring system is to diagnose leaks, obstructions or sensor faults as fast as feasible [24]. Concerning leaks, these may lead to substantial economic damages, harm to the environs as well as health risks. There exist various techniques for the straight diagnosis of leaks relies on visual or palpable physical diagnosis of the fluid like hardware-based or computational pipeline monitoring techniques. Hardware-based techniques are highly based on the physical material installed through the pipeline. However, computational pipeline monitoring techniques are dependent on mathematical models of the pipeline. These techniques are complemented with measuring data of several physical variables related to the flow procedure e.g. pressure, flow rate, and temperature, between others. In the leak detection, apart from the detection of the leak, it is crucial to locate it as precise as feasible. Therefore, it is essential to implement algorithms, which can accurately diagnose the location of the leak since it is not all the time observable from outside the pipe.

The most current published researches are based on designing the observer by using the nonlinear filter or fault diagnosis techniques [36–43]. The functional form of the optimal nonlinear filter has long been recognized, and usual techniques to

S. Razvarz et al., *Flow Modelling and Control in Pipeline Systems*, Studies in Systems, Decision and Control 321, https://doi.org/10.1007/978-3-030-59246-2_8

executable algorithms are divided into two large groups [44]. The first category contains algorithms, which estimate the nonlinear dynamical model and the second category contains algorithms, which estimate the latter distribution numerically, like the point mass filter as well as particle filter [45]. The most popular examples for the first category are the extended Kalman filter (EKF) [46], the unscented Kalman filter (UKF) [47, 48] and related sigma point techniques [49] like the recent quadrature Kalman filter (QKF) as well as cubature Kalman filter (CKF) [50]. They all present the latter filtering distribution accompanied by a state approximation as well as an associated covariance matrix. The standard EKF is generally obtained from the first-order Taylor expansion of the state dynamics as well as measuring model. There exists another version of the EKF, which expands this approach to contain also an estimation of the second-order Taylor term [51], that is known in past few decades. We will indicate the resulting algorithm SEKF to identify it from the standard algorithm.

For decreasing the error in the approximation of the system state and to increase the accuracy of EKF, different modified algorithms are proposed. In an unscented Kalman filter (UKF) algorithm is suggested for nonlinear systems. The algorithm proposed in [52–54] contains the equal computation capacity like the EKF algorithm and also decreases the error in the linearizing nonlinear model. In [55] an algorithm named central difference Kalman filter (CDKF) is proposed and the recursive formula of the algorithm is deduced.

In this paper, a new technique is suggested for analyzing one-dimensional modeling of transients in a pipeline, commonly used for recognition and location of leakage utilizing model-based methods and SEKF observer. The SEKF algorithm is proposed for the nonlinear state-space model, which is based on second-order Taylor expansion of structural functions of the nonlinear system and applies a linearization technique to estimate the quadratic of second-order Taylor expansion of structural functions.

8.2 Pipeline Modeling

In this work, we neglect the convective variation in velocity and the compressibility in the line of length (L). The liquid density (ρ), the flow rate ($\mathbb{Q}$), as well as pressure ($\wp$) at the inlet and outlet of the pipeline are computable for assessment. The cross-sectional area $\left(\text{Å}\right)$ of the pipe is taken to be steady overall the pipe. The proposed pipeline is demonstrated in Fig. 8.1.

The dynamics of fluid in the pipeline is described based on the mass and momentum, as well as the conservation [56]. For obtaining the momentum equation by applying Newton's second law ($F = ma$) to a control volume in the continuum and body force pipe ($s = \frac{\Finv}{2\eth} v$) the below formula is acquired [57],

$$\frac{\partial v}{\partial t} + \frac{1}{\rho}\frac{\partial \wp}{\partial x} + \frac{\Finv}{2\eth} v^2 = 0 \tag{8.1}$$

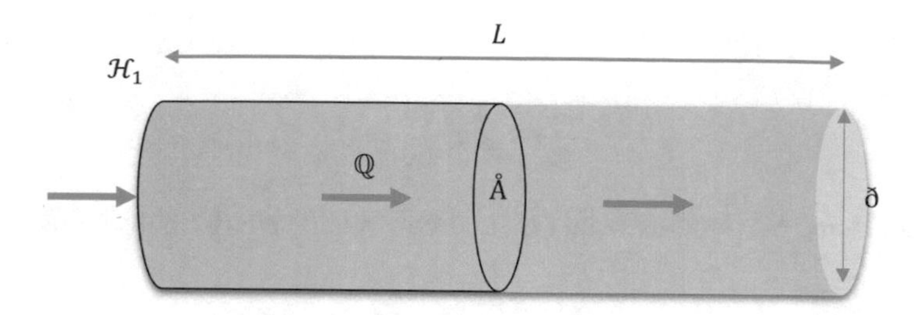

Fig. 8.1 The proposed pipeline system

By replacing $\upsilon = \frac{\mathbb{Q}}{Å}$ and $\wp = \rho g \mathcal{H}$ in (8.1) the below formula is obtained,

$$\frac{\partial \mathbb{Q}}{Å \partial t} + g \frac{\partial}{\partial x} \mathcal{H} + \frac{Ⅎ \mathbb{Q}^2}{2ðÅ^2} = 0 \tag{8.2}$$

Therefore,

$$\frac{\partial \mathbb{Q}}{\partial t} + Åg \frac{\partial}{\partial x} \mathcal{H} + \frac{Ⅎ}{2ð} \frac{\mathbb{Q}^2}{Å} = 0 \tag{8.3}$$

such that $\mathcal{H}$ is the pressure head, $\mathbb{Q}$ is the flow rate, x is the length coordinate, t is the time coordinate, g is the gravity, Å is the section are ð is the diameter, and Ⅎ is the friction coefficient.

Generally, the friction coefficient is taken to be constant, although it is occasionally updated known to rely on the Reynolds number (Re) and the roughness friction coefficient of the pipe (e). The Swamee-Jain equation explains this friction coefficient value for a pipe with a circular section of diameter (ð) as below [58],

$$\mathrm{f} = \left(\frac{0.5}{\ln\left[0.27\left(\frac{e}{ð}\right) + 5.74 \frac{1}{Re^{0.9}}\right]} \right)^2 \tag{8.4}$$

such that the Reynolds number is computed as follows:

$$Re = 4 \frac{\rho \mathbb{Q}}{\pi ð \mu} = \frac{\rho \upsilon ð}{\mu} \tag{8.5}$$

where ρ is taken to be the fluid density also μ is taken to be the fluid viscosity. For $10^{-8} < \frac{e}{ð} < 0.01$ and $5000 < Re < 10^8$, is valid.

In order to obtain the continuity equation, by implementing the law of conservation of mass and also utilizing the Reynolds transport theorem to a control volume and simplifying, the below formula is acquired,

$$\frac{\partial \wp}{\partial t} + \rho a^2 \frac{\partial \nu}{\partial x} = 0 \tag{8.6}$$

By replacing the pressure head ($\mathcal{H}$) and flow rate ($\mathbb{Q}$) in (8.6), the below formula is obtained,

$$\frac{\partial \mathcal{H}}{\partial t} + \frac{a^2}{g\mathring{A}} \frac{\partial \mathbb{Q}}{\partial x} = 0 \tag{8.7}$$

such that a is taken to be the velocity of the wave in an elastic conduit loaded with some compressible fluid. The velocity of the wave can be obtained from the elastic properties of the fluid as well as the pipe. The pressure head ($\mathcal{H}$) and flow rate ($\mathbb{Q}$) are taken to be functions of position and time as $\mathcal{H}(x, t)$ and $\mathbb{Q}(x, t)$, respectively, such that, $x \in [0,\ \mathrm{L}]$, also, L is taken to be the length of the pipe.

For a system having minor variations in the flow rate $\mathbb{Q}$, the momentum formula from the nonlinear system is linearized as follows:

$$\frac{\partial \mathbb{Q}}{\partial t} + \mathring{A}g \frac{\partial}{\partial x} \mathcal{H} + \frac{\Finv}{2\eth} \frac{\mathbb{Q}^2}{\mathring{A}} = 0 \tag{8.8}$$

Using (8.7) and (8.8) we can design the pipeline model. However, acquiring the analytical solutions related to these equations is difficult. Different techniques are proposed for obtaining the numerical solutions for these equations such as characteristics and finite difference methods [59]. In this paper, we apply the finite difference technique.

For converting (8.7) and (8.8) into a system of ordinary differential equations, (8.7) and (8.8) are discretized in the spatial variable. The finite difference approach is considered as a discretization method that partitions the whole pipeline into N number of sections [60]. The finite difference method with a constant step size Δs is considered as the most common discretization technique in model-based techniques for real-time applications. In this paper, we use finite difference technique as it is easy to implement and is a proper technique for nonlinear observer design. The finite difference technique in this paper will be used as below,

$$\frac{\partial \mathbb{Q}(s_{i-1}, t)}{\partial s} \approx \frac{\Delta \mathbb{Q}(s_{i-1}, t)}{\Delta s} \approx \frac{\mathbb{Q}_i - \mathbb{Q}_{i-1}}{\Delta s} \tag{8.9}$$

$$\frac{\partial \mathcal{H}(s_{i-1}, t)}{\partial s} \approx \frac{\Delta \mathcal{H}(s_{i-1}, t)}{\Delta s} \approx \frac{\mathcal{H}_i - \mathcal{H}_{i-1}}{\Delta s} \tag{8.10}$$

$\forall i = 1, \ldots, n$, where n is the number of sections and $\Delta s = s_{i+1} - s_i$ demonstrates the length of the i-section among two successive points of positions. The continuous interval $z \in [0, \mathrm{L}]$ is divided into a three-point discrete part $\{s_k\}$:=$\{0, s_{leak}, \mathrm{L}\}$, such that z_{leak} demonstrates the indistinct leak position that shows in Fig. 8.2. The flow

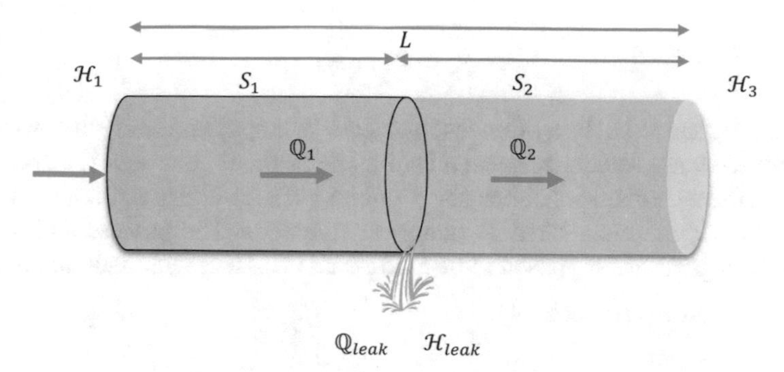

Fig. 8.2 Pipeline with leakage scheme

rate of the leak is estimated with $\mathbb{Q}_{leak} = C_d \mathring{A}_{leak}\sqrt{2g}\sqrt{\mathcal{H}(s_{leak}, t)}$, where C_d is the discharge coefficient, also $\mathring{A}_{leak}$ is the leak cross-section area. The leakage flow rate is obtained as $\mathbb{Q}_{leak} = \Lambda\sqrt{\mathcal{H}(s_{leak}, t)}$, where $\Lambda = C_d \mathring{A}_{leak}\sqrt{2g}$. The dynamic model of the pipeline system is stated by an ordinary differential equation system as follows:

$$
\begin{aligned}
\dot{\mathbb{Q}}_1 &= \frac{g\mathring{A}}{s}(\mathcal{H}_1 - \mathcal{H}_2) - \frac{\Finv}{2\eth}\frac{\mathbb{Q}_1^2}{\mathring{A}} \\
\dot{\mathcal{H}}_{leak} &= \frac{c^2}{g\mathring{A}s}\left(\mathbb{Q}_1 - \mathbb{Q}_2 - \Lambda\sqrt{\mathcal{H}_{leak}}\right) \\
\dot{\mathbb{Q}}_2 &= \frac{g\mathring{A}}{\mathrm{L} - s}(\mathcal{H}_2 - \mathcal{H}_3) - \frac{\Finv}{2\eth}\frac{\mathbb{Q}_2^2}{\mathring{A}}
\end{aligned}
\tag{8.11}
$$

It has been assumed that the inlet pressure h_1 and the outlet pressure h_3 are known, which are externally defined by the pump power. The pressure h_2 at the leak point as well as the inlet and outlet flow rates ($\mathbb{Q}_1$ and $\mathbb{Q}_2$) is taken to be the indistinct dynamic variable. According to the continuity equation in pipeline we have,

$$
\mathbb{Q}_1 = \mathbb{Q}_{leak} + \mathbb{Q}_2 \tag{8.12}
$$

8.3 Observer Design

8.3.1 Nonlinear State Space Model

We will use filtering algorithms for the following nonlinear state space model,

$$
x_{i+1} = \varphi(x_i, u_i) \tag{8.13}
$$

$$y_k = \zeta(x_i, e_i) \tag{8.14}$$

where, x_i is taken to be the n_x dimensional state vector, u_i and e_i are the process and measurement noise vectors, respectively (dimensions $n_u \times 1$ and $n_e \times 1$, respectively), also, y_k is the n_y-dimensional measured output vector. State transition φ as well as measurement ζ are considered as nonlinear functions. The process and measurement noise signals are supposed to be zero-mean Gaussian with definite covariance matrices $\breve{\mathbb{Q}}_i$ and $\breve{R}_i$, respectively.

8.3.2 System Approximation by Taylor Expansion

The nonlinear state, as well as the measurement equation in (8.13), can be estimated using Taylor expansion around a specified state and a nominal (zero) noise value. Nevertheless, although derivative terms greater than order two are ignored in the expansion, the expressions are complicated. Therefore, we introduce new variables, which contain state and noise.

Defining $\mu_i = \left(x_i^T, u_i^T\right)$, (8.13) is rewritten when $x_{i+1} = \varphi(\mu_i)$. A Taylor expansion for each scalar element x_{i+1}^l of the state vector x_{i+1} around a known $\hat{\mu}_i$ is stated as below

$$\begin{aligned} x_{i+1}^l &\approx \varphi_l(\hat{\mu}_i) + \left(\nabla_\mu \varphi_l(\hat{\mu}_i)\right)^T (\mu_i - \hat{\mu}_i) \\ &+ \frac{1}{2}(\mu_i - \hat{\mu}_i)^T \nabla_\mu^2 \varphi_l(\hat{\mu}_i)(\mu_i - \hat{\mu}_i) \\ &l \in [1, \ldots, n_x] \end{aligned} \tag{8.15}$$

such that $\left(\nabla_\mu \varphi_l(\hat{\mu}_i)\right)^T$ is taken to be a row vector having the partial derivatives of $\varphi_l(\mu_i)$ regarding all elements of μ_i measured at $\hat{\mu}_i$. Consequently, $\nabla_\mu^2 \varphi_l(\hat{\mu}_i)$ is a Hessian matrix having second order partial derivatives. Similarly, (8.14) is rewritten when $y_{i+1} = \zeta(\epsilon_i)$. A component of y_i is defined as below,

$$\begin{aligned} y_i^m &\approx \zeta_m(\hat{\epsilon}_i) + \left(\nabla_\epsilon \zeta_m(\hat{\epsilon}_i)\right)^T (\epsilon_i - \hat{\epsilon}_i) \\ &+ \frac{1}{2}(\epsilon_i - \hat{\epsilon}_i)^T \nabla_\mu^2 \zeta_m(\hat{\epsilon}_i)(\epsilon_i - \hat{\epsilon}_i) \\ &m \in \left[1, \ldots, n_y\right] \end{aligned} \tag{8.16}$$

The represented first as well as second order derivatives, for example, $\left(\nabla_\mu \varphi_l(\hat{\mu}_i)\right)^T$ or $\nabla_\mu^2 \zeta_m(\hat{\epsilon}_i)$ are concatenated vectors and block matrices of derivatives with regard to x, and u or e. Furthermore, mixed second order derivatives happen in the Hessians. Stacking, for example, all n_y rows $\left(\nabla_\epsilon \zeta_m(\hat{\epsilon}_i)\right)^T$ generates the Jacobian matrix $\left(\nabla_x \zeta_m(\hat{\epsilon}_i)\right)^T$ that will be utilized in the Kalman gain in the following subsection.

8.3.3 *Second Order Extended Kalman Filter Recursions*

Defining the initial state x_0 as well as its covariance matrix P_0, the filtering algorithm contains recursively named the time and measurement updates.

8.3.3.1 Time Update

A predicted approximation $\hat{x}^l_{i+1|i}$ of $x^l_{i+1|i}$ is obtained by implementing the expectation of as

$$\hat{x}^l_{i+1|i} = \varphi_l(\hat{\mu}_{i|i}) + \frac{1}{2} tr\left(\nabla^2_x \varphi_l(\hat{\mu}_{i|i}) \breve{P}_{i|i}\right) + \frac{1}{2} tr\left(\nabla^2_u \varphi_l(\hat{\mu}_{i|i}) \breve{\mathbb{Q}}_i\right) \tag{8.17}$$

It is supposed that $\hat{\mu}_{i|i} = \left(\hat{x}^T_{i|i}, 0^T\right)^T = E\left(\left(x^T_i, u^T_i\right)^T | y_i\right)$. Therefore, odd terms in μ_i disappear while executing the expectation. Moreover, the connection among u_i and x_i is not considered that can be stated by covariance matrix $(\mu_i | y_i) =$ Block diagonal matrix $\left(\breve{P}_{i|i}, \breve{\mathbb{Q}}_i\right)$, hence no mixed second order derivatives happen in (8.17) [61, 62]. The predicted covariance matrix $P_{i+1|i} =$ covariance matrix $(x_{i+1} | y_i)$ is defined by its elements in row l and column m as below,

$$\begin{aligned} \breve{P}^{lm}_{i+1|i} &= \left(\nabla_x \varphi_l(\hat{\mu}_{i|i})\right)^T \breve{P}_{i|i} \nabla_x \varphi_m(\hat{\mu}_{i|i}) \\ &+ \left(\nabla_u \varphi_l(\hat{\mu}_{i|i})\right)^T \breve{\mathbb{Q}}_i \nabla_u \varphi_m(\hat{\mu}_{i|i}) \\ &+ \frac{1}{2} tr\left(\nabla^2_x \varphi_l(\hat{\mu}_{i|i}) \breve{P}_{i|i} \nabla^2_x \varphi_m(\hat{\mu}_{i|i}) \breve{P}_{i|i}\right) \\ &+ \frac{1}{2} tr\left(\nabla^2_u \varphi_l(\hat{\mu}_{i|i}) \breve{\mathbb{Q}}_i \nabla^2_u \varphi_m(\hat{\mu}_{i|i}) \breve{\mathbb{Q}}_i\right) \end{aligned} \tag{8.18}$$

such that $l, m \in [1, \ldots, n_x]$. The covariance calculation is simple, however needs a Gaussian presumption on x_i.

8.3.3.2 Measurement Update

Carrying out the expectation of regarding a state prediction generates an output approximation as follows:

$$\hat{y}^{m}_{i|i-1} = \zeta_m\left(\hat{\epsilon}_{i|i-1}\right) + \frac{1}{2}tr\left(\nabla^2_x \zeta_m\left(\hat{\epsilon}_{i|i}\right)\breve{P}_{i|i-1}\right) + \frac{1}{2}tr\left(\nabla^2_e \zeta_m\left(\hat{\epsilon}_{i|i}\right)\breve{R}_i\right) \tag{8.19}$$

The output covariance matrix $\breve{S}_k =$ covariance matrix $(y_k|y_{k-1})$ is defined by its elements $l, m \in \left[1, \ldots, n_y\right]$ as follows:

$$\begin{aligned}\breve{S}^{lm}_i &= \left(\nabla_x \zeta_l\left(\hat{\epsilon}_{i|i-1}\right)\right)^T \breve{P}_{i|i-1} \nabla_x \zeta_m\left(\hat{\epsilon}_{i|i-1}\right) \\ &+ \left(\nabla_e \zeta_l\left(\hat{\epsilon}_{i|i-1}\right)\right)^T \breve{R}_i \nabla_e \zeta_m\left(\hat{\epsilon}_{i|i-1}\right) \\ &+ \frac{1}{2}tr\left(\nabla^2_x \zeta_l\left(\hat{\epsilon}_{i|i-1}\right)\breve{P}_{i|i-1} \nabla^2_x \zeta_m\left(\hat{\epsilon}_{i|i-1}\right)\breve{P}_{i|i-1}\right) \\ &+ \frac{1}{2}tr\left(\nabla^2_e \varphi_l\left(\hat{\epsilon}_{i|i-1}\right)\breve{R}_i \nabla^2_e \zeta_m\left(\hat{\epsilon}_{i|i-1}\right)\breve{R}_i\right)\end{aligned} \tag{8.20}$$

The uncorrelated noise has been supposed as $cov(\epsilon_i|y_{i-1}) = blkdiag\left(\breve{P}_{i|i-1}, \breve{R}_i\right)$. The Kalman gain is acquired under the supposition that x_i and y_i are jointly Gaussian distributed as stated by

$$\begin{pmatrix} x_i \\ y_i \end{pmatrix} \sim N\left(\begin{pmatrix} \hat{x}_{i|i-1} \\ \hat{y}_{i|i-1} \end{pmatrix}, \begin{pmatrix} \breve{P}_{i|i-1} & \breve{M}_i \\ \breve{M}^T_i & \breve{S}_i \end{pmatrix}\right) \tag{8.21}$$

where the cross terms are defined as $\breve{M}_i =$ covariance matrix $(x_i, y_i|y_{i-1}) = \breve{P}_{i|i-1} \nabla_x \zeta\left(\hat{\epsilon}_{i|i-1}\right)$. Applying optimal estimation theory [51], the posterior approximation is obtained as conditional expectation of x_i as below,

$$\hat{x}_{i|i} = \hat{x}_{i|i-1} + \breve{M}_i \breve{S}^{-1}_i \left(y_i - \hat{y}_{i|i-1}\right) \tag{8.22}$$

The iteration is finalized by updating the covariance matrix

$$\breve{P}_{i|i} = \breve{P}_{i|i-1} + \breve{M}_i \breve{S}^{-1}_i \breve{M}^T_i \tag{8.23}$$

Finally, $\breve{M}$ and $\breve{S}$ can be chosen as the covariance matrices of measure and process noises, respectively, or just as tuning parameters, with $\breve{P} = \breve{P}^{T} > 0$, $\breve{M} = \breve{M}^{T} > 0$ and $\breve{S} = \breve{S}^{T} > 0$.

8.4 Simulation Results

In this section, simulations results are reported for one case.

The simulations are carried out by Matlab environment. The step time of the solver (ODE3) is set to $\Delta t = 0.001$. An illustration of the model is shown in Fig. 8.1. The model has been realized on the pipeline with the following physical parameters:

The length of the pipeline is $L = 120$ m, the diameter of the pipe is $\eth = 0.08$ m, the cross-section is $\text{Å} = 5.03 \times 10^{-3}\,\text{m}^2$, density is $\rho = 1000\,\text{kg/m}^3$, gravity is $g = 9.81\,\text{m/s}^2$, the friction factor of the pipe is $f = 0.06$, the friction factor of and the wave speed is $a = 1250\,\text{m/s}$.

The covariance matrices of measure and process noises introduced by expressions respectively

$$\breve{M} = \text{Diagonal Matrix}\left[3.5 \times 10^{-5}, 5 \times 10^{-2}, 3.5 \times 10^{-5}, 1000, 10^{-7}\right]$$

$$\breve{S} = \text{Diagonal Matrix}\left[3.5 \times 10^{-5}, 3.5 \times 10^{-5}\right]$$

The initialization of the state is given in Table 8.1.

The pressure heads are shown in Fig. 8.3, and the flow rate are shown in Fig. 8.4 where it can be seen the effect of first leak at time $t = 30$ s.

The corresponding detection and isolation results are shown in Figs. 8.5, 8.6 and 8.7. Figure 8.5 are shown, the measured pressure heads that correspond to leak points and shown, validate results obtained by observer.

In Fig. 8.6, is shown the estimated leak location. In Fig. 8.7 the estimation leak coefficient is shown. In this case the leak identification is pretty successful.

Table 8.1 Initialization SEKF

CEKF	Value	Units
$\hat{\mathbb{Q}}_1$	7.75×10^{-3}	(m^3/s)
$\hat{\mathcal{H}}_{leak}$	8.32	(m)
$\hat{\mathbb{Q}}_2$	7.75×10^{-3}	(m^3/s)
$\hat{s}$	45.42	(m)
$\hat{\Lambda}$	0	$\left(\text{m}^{\frac{5}{2}}/\text{s}\right)$

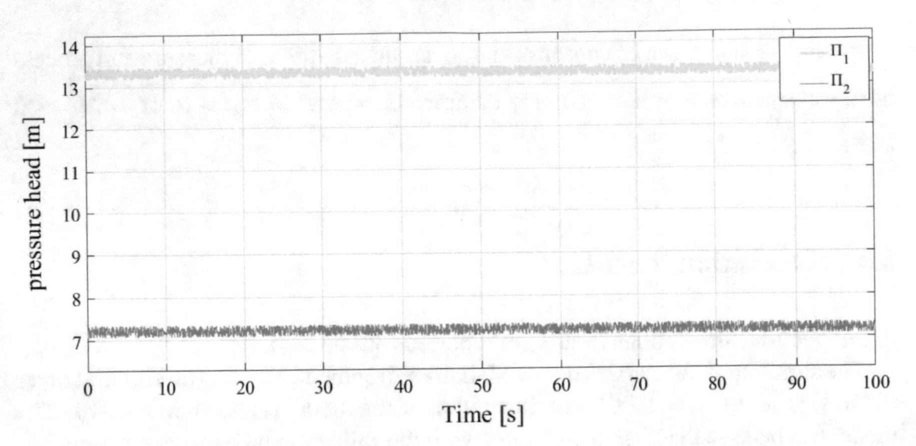

Fig. 8.3 Pressure heads at the pipeline

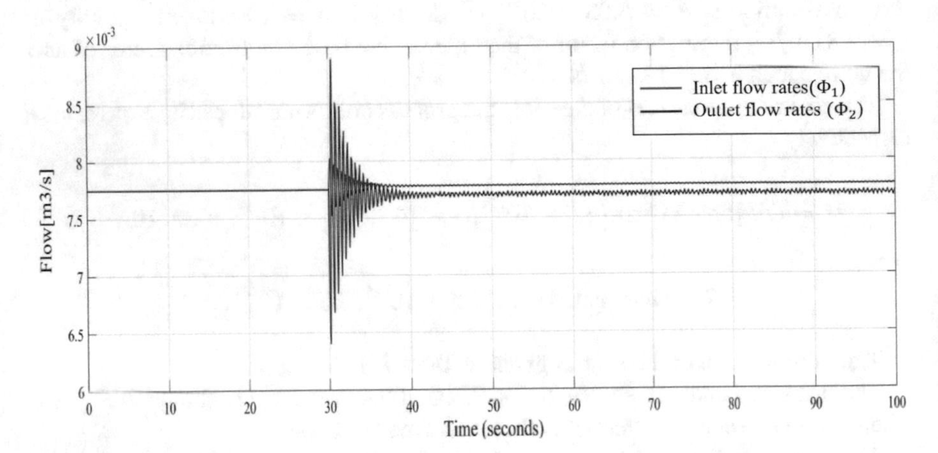

Fig. 8.4 The flow rate at the pipeline

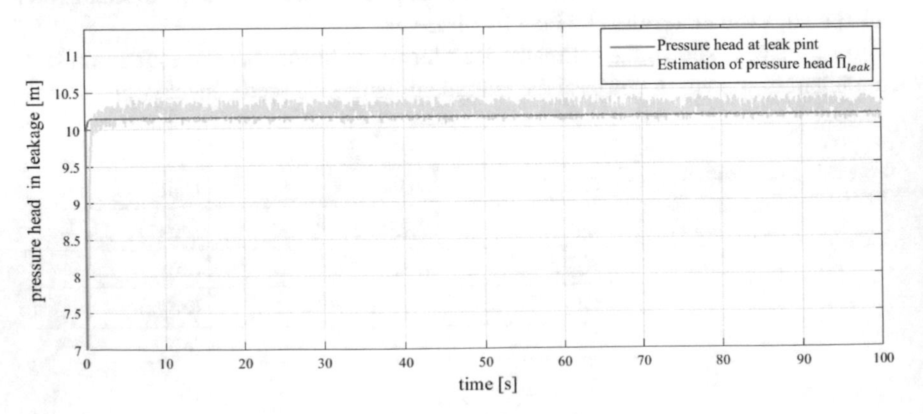

Fig. 8.5 Estimation of pressure head $\dot{\mathcal{H}}_{leak}$ and measurements of pressure head at leak point

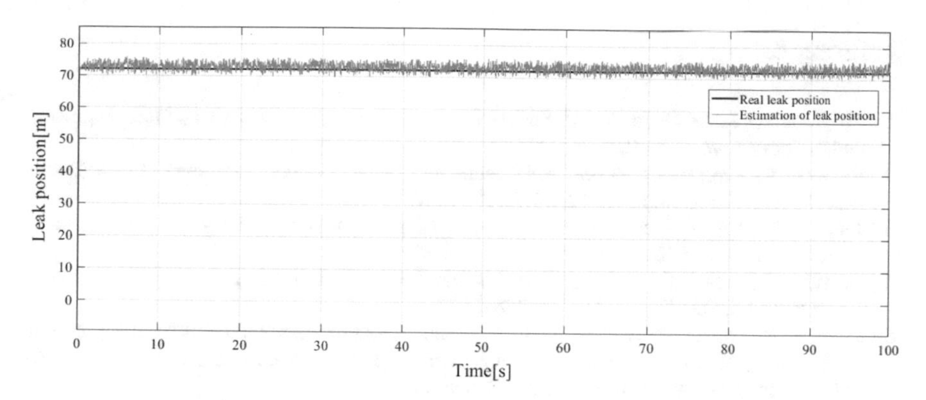

Fig. 8.6 Estimation and real leak position

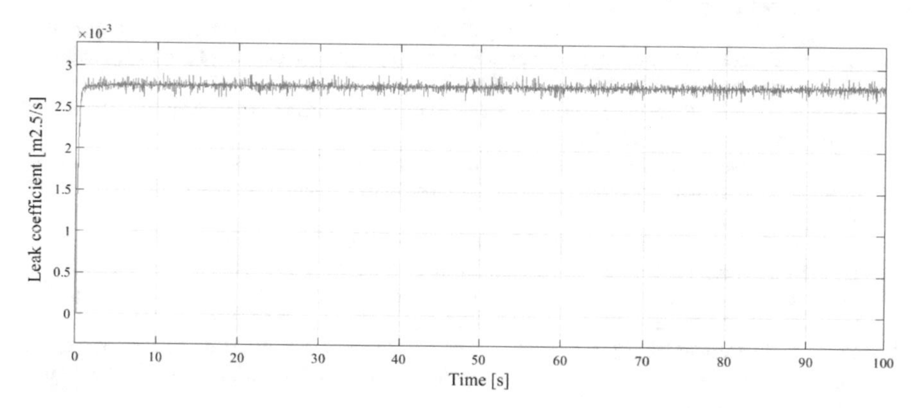

Fig. 8.7 Estimation of leak coefficients $\hat{\Lambda}$

8.5 Conclusions

In this paper, a technique based on SEKF is proposed to detect leakage in the pipeline. The SEKF algorithm is proposed for the nonlinear state-space model, which is based on second-order Taylor expansion of structural functions of the nonlinear system and applies a linearization technique to estimate the quadratic of second-order Taylor expansion of structural functions. A simulation example is given for demonstrating the effectiveness of the proposed technique. In future research, the observability, as well as the controllability of the pipeline system will be investigated.

References

1. Yu, W., Jafari, R.: Modeling and Control of Uncertain Nonlinear Systems with Fuzzy Equations and Z-Number. Wiley (2019)
2. Razvarz, S., Jafari, R.: ICA and ANN modeling for photocatalytic removal of pollution in wastewater. Math. Comput. Appl. **22**(3), 38 (2017)
3. Jafari, R., Razvarz, S., Gegov, A.: Neural network approach to solving fuzzy nonlinear equations using Z-numbers. IEEE Trans. Fuzzy Syst. (2019)
4. Razvarz, S., Jafari, R.: Intelligent techniques for photocatalytic removal of pollution in wastewater. J. Electr. Eng. **5**(1), 321–328 (2017)
5. Jafari, R., Razvarz, S., Gegov, A., Paul, S., Keshtkar, S.: Fuzzy Sumudu transform approach to solving fuzzy differential equations with Z-numbers. In: Advanced Fuzzy Logic Approaches in Engineering Science, pp. 18–48. IGI Global (2019)
6. Jafari, R., Razvarz, S., Gegov, A.: A novel technique to solve fully fuzzy nonlinear matrix equations. In: International Conference on Theory and Applications of Fuzzy Systems and Soft Computing, pp. 886–892. Springer (2018)
7. Jafari, R., Razvarz, S., Gegov, A.: Fuzzy differential equations for modeling and control of fuzzy systems. In: International Conference on Theory and Applications of Fuzzy Systems and Soft Computing, pp. 732–740. Springer (2018)
8. Jafari, R., Yu, W., Razvarz, S., Gegov, A.: Numerical methods for solving fuzzy equations: a survey. Fuzzy Sets Syst. (2019)
9. Jafari, R., Razvarz, S., Gegov, A.: A new computational method for solving fully fuzzy nonlinear systems. In: International Conference on Computational Collective Intelligence, pp. 503–512. Springer (2018)
10. Jafari, R., Razvarz, S., Gegov, A., Paul, S.: Modeling and control of uncertain nonlinear systems. In: 2018 International Conference on Intelligent Systems (IS), pp. 168–173. IEEE (2018)
11. Jafari, R., Razvarz, S., Gegov, A.: A novel technique for solving fully fuzzy nonlinear systems based on neural networks. Vietnam J. Comput. Sci. **7**(1), 93–107 (2020)
12. Razvarz, S., Hernández-Rodríguez, F., Jafari, R., Gegov, A.: Foundation of Z-numbers and engineering applications. In: Latin American Symposium on Industrial and Robotic Systems, pp. 15–24. Springer (2019)
13. Jafari, R., Contreras, M.A., Yu, W., Gegov, A.: Applications of fuzzy logic, artificial neural network and neuro-fuzzy in industrial engineering. In: Latin American Symposium on Industrial and Robotic Systems, pp. 9–14. Springer (2019)
14. Jafari, R., Razvarz, S., Gegov, A., Yu, W.: Fuzzy control of uncertain nonlinear systems with numerical techniques: a survey. In: UK Workshop on Computational Intelligence, pp. 3–14. Springer (2019)
15. Jafari, R., Razvarz, S., Yu, W., Gegov, A., Goodwin, M., Adda, M.: Genetic algorithm modeling for photocatalytic elimination of impurity in wastewater. In: Proceedings of SAI Intelligent Systems Conference, pp. 228–236. Springer (2019)
16. Tatchum, M., Gegov, A., Jafari, R., Razvarz, S.: Parallel distributed compensation for voltage controlled active magnetic bearing system using integral fuzzy model. In: 2018 International Conference on Intelligent Systems (IS), pp. 190–198. IEEE (2018)
17. Razvarz, S., Jafari, R., Gegov, A.: Solving partial differential equations with Bernstein neural networks. In: UK Workshop on Computational Intelligence, pp. 57–70. Springer (2018)
18. Jafarian, A., Jafari, R.: New iterative approach for solving fully fuzzy polynomials. Int. J. Fuzzy Math. Syst. **3**(2), 75–83
19. Jafarian, A., Jafari, R.: New method for solving fuzzy polynomials. Adv. Fuzzy Math. **8**(1), 25–33 (2013)
20. Jafarian, A., Jafari, R.: An iterative method for solving fuzzy polynomials by fuzzy neural networks (2012)
21. Jafarian, A., Jafari, R.: Simulation and evaluation of fuzzy polynomials by feed-back neural networks (2012)

22. Jafari, R., Yu, W.: Fuzzy control for uncertainty nonlinear systems with dual fuzzy equations. J. Intell. Fuzzy Syst. **29**(3), 1229–1240 (2015)
23. Jafari, R., Yu, W.: Fuzzy modeling for uncertainty nonlinear systems with fuzzy equations. Math. Probl. Eng. (2017)
24. Verde, C., Torres, L.: Modeling and Monitoring of Pipelines and Networks. Springer (2017)
25. Verde, C., Visairo, N., Gentil, S.: Two leaks isolation in a pipeline by transient response. Adv. Water Resour. **30**(8), 1711–1721 (2007)
26. Verde, C., Visairo, N.: Identificability of multi-leaks in a pipeline. In: Proceedings of the 2004 American Control Conference. IEEE (2004)
27. Verde, C., Torres, L., González, O.: Decentralized scheme for leaks' location in a branched pipeline. J. Loss Prev. Process Ind. **43**, 18–28 (2016)
28. Besançon, G., Georges, D., Begovich, O., Verde, C., Aldana, C.: Direct observer design for leak detection and estimation in pipelines. In: 2007 European Control Conference (ECC), pp. 5666–5670. IEEE (2007)
29. Verde, C.: Accommodation of multi-leak location in a pipeline. Control Eng. Pract. **13**(8), 1071–1078 (2005)
30. Verde, C.: Minimal order nonlinear observer for leak detection. J. Dyn. Syst. Meas. Control **126**(3), 467–472 (2004)
31. Carrera, R., Verde, C., Cayetano, R.: A SCADA expansion for leak detection in a pipeline. Sensors **2300**(2320), 2340 (2015)
32. Verde, C.: Leakage location in pipelines by minimal order nonlinear observer. In: Proceedings of the 2001 American Control Conference (Cat. No. 01CH37148), pp. 1733–1738. IEEE (2001)
33. Verde, C., Sánchez-Parra, M.: Application of structural analysis to improve fault diagnosis in a gas turbine. In: Gas Turbines, p. 307 (2010)
34. Kowalczuk, Z., Gunawickrama, K.: Leak detection and isolation for transmission pipelines via nonlinear state estimation. IFAC Proc. Vol. **33**(11), 921–926 (2000)
35. Garcia, J., Leon, B., Begovich, O.: Validation of a semiphysical pipeline model for multi-leak diagnosis purposes. In: Proceedings of the 20th IASTED International Conference, vol. 060, p. 24 (2009)
36. Razvarz, S., Vargas-Jarillo, C., Jafari, R.: Pipeline monitoring architecture based on observability and controllability analysis. In: 2019 IEEE International Conference on Mechatronics (ICM), 18–20 Mar 2019, pp. 420–423 (2019)
37. Jafari, R., Razvarz, S., Vargas-Jarillo, C., Yu, W.: Control of flow rate in pipeline using PID controller. In: 2019 IEEE 16th International Conference on Networking, Sensing and Control (ICNSC), 9–11 May 2019, pp. 293–298 (2019)
38. Jafari, R., Razvarz, S., Vargas-Jarillo, C., Gegov, A.: Blockage detection in pipeline based on the extended Kalman filter observer. Electronics **9**(1), 91–107 (2020)
39. Razvarz, S., Jafari, R., Vargas-Jarillo, C.: Modelling and analysis of flow rate and pressure head in pipelines. In: 2019 16th International Conference on Electrical Engineering, Computing Science and Automatic Control (CCE), pp. 1–6. IEEE (2019)
40. Jafari, R., Razvarz, S., Vargas-Jarillo, C., Gegov, A.E.: The effect of baffles on heat transfer. In: ICINCO (2), pp. 607–612 (2019)
41. Razvarz, S., Jafari, R., Vargas-Jarillo, C., Gegov, A., Forooshani, M.: Leakage detection in pipeline based on second order extended Kalman filter observer. IFAC-PapersOnLine **52**(29), 116–121 (2019)
42. Razvarz, S., Vargas-Jarillo, C., Jafari, R., Gegov, A.: Flow control of fluid in pipelines using PID controller. IEEE Access **7**, 25673–25680 (2019)
43. Razvarz, S., Chavez, L.F.G., Vargas-Jarillo, C.: Nanotechnology applications in industry and heat transfer. In: Latin American Symposium on Industrial and Robotic Systems, pp. 1–8. Springer (2019)
44. Hoteit, I., Pham, D.T., Triantafyllou, G., Korres, G.: A new approximate solution of the optimal nonlinear filter for data assimilation in meteorology and oceanography. Mon. Weather Rev. **136**(1), 317–334 (2008)

45. Gustafsson, F.: Particle filter theory and practice with positioning applications. IEEE Aerosp. Electron. Syst. Mag. **25**(7), 53–82 (2010)
46. Athans, M., Wishner, R., Bertolini, A.: Suboptimal state estimation for continuous-time nonlinear systems from discrete noisy measurements. IEEE Trans. Autom. Control **13**(5), 504–514 (1968)
47. Julier, S.J., Uhlmann, J.K., Durrant-Whyte, H.F.: A new approach for filtering nonlinear systems. In: Proceedings of 1995 American Control Conference-ACC'95, pp. 1628–1632. IEEE (1995)
48. Wan, E.A., Van Der Merwe, R.: The unscented Kalman filter for nonlinear estimation. In: Proceedings of the IEEE 2000 Adaptive Systems for Signal Processing, Communications, and Control Symposium (Cat. No. 00EX373), pp. 153–158. IEEE (2000)
49. Quine, B.M.: A derivative-free implementation of the extended Kalman filter. Automatica **42**(11), 1927–1934 (2006)
50. Arasaratnam, I., Haykin, S.: Cubature Kalman filters. IEEE Trans. Autom. Control **54**(6), 1254–1269 (2009)
51. Anderson, B.D., Moore, J.B.: Optimal Filtering. Courier Corporation (2012)
52. Lee, J.H., Ricker, N.L.: Extended Kalman filter based nonlinear model predictive control. Ind. Eng. Chem. Res. **33**(6), 1530–1541 (1994)
53. Schei, T.S.: A finite-difference method for linearization in nonlinear estimation algorithm (1998)
54. Haus, B., Aschemann, H., Mercorelli, P.: Tracking control of a piezo-hydraulic actuator using input–output linearization and a cascaded extended Kalman filter structure. J. Franklin Inst. **355**(18), 9298–9320 (2018)
55. Zhen, Q., Yingjie, Q.: A SLAM algorithm based on an iterated central difference particle filter. J. Harbin Eng. Univ. **3**, 18 (2012)
56. Darcy, H., Law, H.: The History of the Darcy-Weisbach Equation
57. Brown, G.O.: The history of the Darcy-Weisbach equation for pipe flow resistance. In: Environmental and Water Resources History, pp. 34–43 (2003)
58. Swanee, P., Jain, A.K.: Explicit equations for pipeflow problems. J. Hydraul. Div. **102**(5) (1976)
59. Wylie, E.B., Streeter, V.L.: Fluid Transients. MHI (1978)
60. Beilina, L.: Domain decomposition finite element/finite difference method for the conductivity reconstruction in a hyperbolic equation. Commun. Nonlinear Sci. Numer. Simul. **37**, 222–237 (2016)
61. Hendeby, G.: Performance and Implementation Aspects of Nonlinear Filtering. Institutionen för systemteknik (2008)
62. Tanizaki, H.: Nonlinear Filters: Estimation and Applications. Springer Science & Business Media (2013)

Chapter 9
Control of Flow Rate in Heavy-Oil Pipelines Using PD and PID Controller

9.1 Introduction

Classic PD and PID approaches as well as controllers are updated and expanded during the years, from the primary controllers on the basis of the relays as well as synchronous electric motors or pneumatic or hydraulic systems to current microprocessors. Currently, many techniques for the tuning as well as design of PI and PID controllers are proposed [1]. The method proposed in [2] is the most widely utilized PID parameter tuning methodology in chemical industry and is considered as a conventional technique. Basilio and Matos [3] suggested a new method with less complexity in order to tune the parameters of PI controllers of the plant with monotonic step response. The methodology of internal mode principle is utilized in [4, 5] in order to extract the gains of PID and PI controllers. Exhaustive investigation [6–8] revealed that the outcomes of P control are very sensitive to the sensing location as well as the quantity of phase shift. By suitable selections of these variables, the P control can be completely efficient in annihilating the vortex shedding or minimizing its strength. Furthermore, it is demonstrated that the increment in the proportional gain can results in the decrement of the velocity fluctuations in the wake and the strength of vortex shedding. Nevertheless, a large gain causes instability in the system [9–11].

In order to implement the control law, the primary step is to determine a desired output response of a particular system to an arbitrary input over a time interval, that can be carried out by system identification [12]. Generally, it is feasible to generate a model on the basis of a complete physical illustration of the system. Nevertheless, this model contains complexity, also has high calculation costs [13]. The secondary step is to define the parameters of the PID controller. There exist various literatures associated with the methodologies for tuning of PID controllers applied in various controller structures [14].

Flow control is a major rapidly evolving field of fluid mechanics. There have been various concepts of flow control in drag reduction, lift enhancement, mixing enhancement, etc. [15–17]. Fadlun et al. [18] implemented the concept of [19] to

S. Razvarz et al., *Flow Modelling and Control in Pipeline Systems*, Studies in Systems, Decision and Control 321, https://doi.org/10.1007/978-3-030-59246-2_9

a finite-difference methodology where a staggered grid is used. In [20] a digital pulse feedback flow control system utilizing microcontroller as well as feedback sensing element is developed. In the recent years artificial neural networks (ANNs) have become popular and many interesting ANN applications have been reported in engineering field [21–43]. In [44] the authors have develop flow modeling and optimization methods using neural networks. In [45] a neural network controller approach is proposed in order to control the pump flow rate. Surprisingly, even though the flow control methods are widely spread, investigating the stability of the control system is very rare. Surprisingly, even though the flow control methods [46–53] are widely spread, investigating the stability of the control system is very rare. In [7, 10, 54] the P, PI and PID controls are proposed for flow over a cylinder with a Reynolds number blow the 200. The aim of control in these studies is the attenuation or annihilation of vortex shedding behind a bluff body. The only investigation on the implementation of PI and PID controls to the flow over a bluff body is carried out in [7].

This chapter deals with the modeling and control of flow rate in heavy-oil pipelines. For this aim, the PD and PID control algorithm is utilized to control the flow mechanism in pipelines. A torsional actuator is placed on the motor-pump in order to control the vibration on motor. The stability of the PD and PID controller is verified using Lyapunov stability analysis. The stability analysis of the controller results in a theorem which validate that the system states are bounded. The theoretical concepts are validated using numerical simulations and analysis, which proves the effectiveness of the PD and PID controller in the control of flow rates in pipelines. Further, it is the first attempt to place a torsional actuator on the motor-pump in order to control the vibration on motor and hence control the flow rates in pipelines.

9.2 Materials and Methods for Modelling of the System

Flow control loop system is basically a feedback control system. The structure of the pump model system is shown in Fig. 9.1 which is an open loop system. If there is unwanted vibration in the motor, the stability of the flow rate will hamper. Therefore, it is important to change the open loop system to a closed loop system by

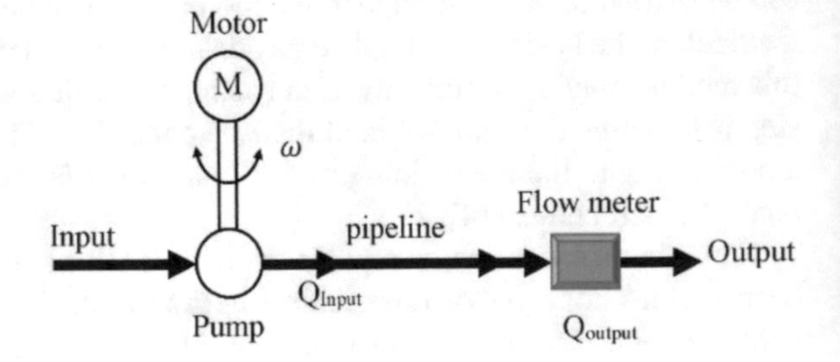

Fig. 9.1 Scheme of open loop model

implementing a controller so as to control the stability of flow rate by controlling the vibration in the motor.

9.2.1 *Modelling of the Pipeline*

The proposed model consists of induction motor, which causes a rotation in pump and consequently can lead to flow of heavy-oil in pipelines as shown in (Fig. 9.1). This flow model can be illustrated in the form of partial differential equation (PDE) [55].

The linear global theory associated with flow stability is rooted in eigen decompositions of the linearized flow operators. The direct as well as adjoint eigen decompositions associated with these kinds of operators generate information related to the stability of the operator, the acceptance of initial conditions as well as external forcing, also the sensitivity to spatially localized disturbances.

From the viewpoint of incompressible, constant-density, constant-viscosity flows associated with Newtonian fluids, the nonlinear Navier–Stokes equation is defined for a nondimensional velocity field $(x, t) : \mathbb{R}^n \times \mathbb{R} \rightarrow \mathbb{R}^n$, pressure field $\wp(x, t) : \mathbb{R}^n \times \mathbb{R} \rightarrow \mathbb{R}^n$ and Reynolds number $Re > 0$ as below equation,

$$\frac{\partial \upsilon}{\partial t} = -\frac{\nabla \wp}{\rho} - \upsilon \cdot \nabla \upsilon + \Finv_f \tag{9.1}$$

where,

ρ is the density in $\frac{\text{kg}}{\text{m}^3}$.

υ is the flow velocity in $\frac{\text{m}}{\text{s}}$,

∇ is the divergence,

$\wp$ is the pressure in $\frac{\text{kg}}{\text{ms}^2}$,

t is time in s,

$\Finv_f$ is termed as the summation of external force and body forces.

By implementing the mass balance into Eq. (9.1) the following is concluded [56],

$$\nabla \cdot \upsilon = 0 \tag{9.2}$$

Equation (9.1) can be rewritten as,

$$\frac{\partial \upsilon}{\partial t} = -\frac{1}{\rho}\frac{\partial \wp}{\partial x} + \Finv_f \tag{9.3}$$

Let $\frac{\partial \wp}{\partial x}$ be the change of pressure in two different points, moreover for achieving a numerical stability of computation, it is essential to partition pipeline into various segment (generally identical), hence the flow in pipeline can be stated as,

$$\frac{\partial \upsilon_i}{\partial t} = -\frac{1}{\rho L}(\wp_i - \wp_{i-1}) + \daleth_f, \quad i = 1, \ldots, n \tag{9.4}$$

where L is taken to be the distance between two sections. Now let,

$$\alpha \wp_i = \wp_{i-1}, \quad i = 1, \ldots, n \tag{9.5}$$

where α is termed as the coefficient of pressure changes in sections

$$\frac{\partial \upsilon_i}{\partial t} = -\frac{1}{\rho L}(1 - \alpha)\wp_i + \daleth_f, \quad i = 1, \ldots, n \tag{9.6}$$

The loss of the friction under the conditions of laminar flow conforms with the Hagen–Poiseuille equation [57, 58]. For a circular pipe having a fluid of density (ρ) and Kinematic viscosity $\tilde{\upsilon}$, the hydraulic slope F_f can be described as,

$$\daleth_f = \frac{64}{Re}\frac{\upsilon^2}{2gD} + F_b = \frac{64\tilde{\upsilon}}{2g}\frac{\upsilon}{D^2} + F_b \tag{9.7}$$

where g is the gravity, D is the diameter of the pipes and F_b is the shape force vector in pipes. By substitution of (9.7) in (9.6), the following equation can be extracted,

$$\frac{\partial \upsilon_i}{\partial t} = -\frac{1}{\rho L}(1 - \alpha)\wp_i + \frac{64\tilde{\upsilon}}{2g}\frac{\upsilon}{D^2} + F_b, \quad i = 1, \ldots, n \tag{9.8}$$

The pressure $\wp(x, t)$ in the pipeline can be described as,

$$\wp = \frac{F}{\AA} \tag{9.9}$$

where F is the force inside the pipeline, also Å is the cross section in the pipe.

By taking into consideration $F = ma = m\frac{d^2x}{d^2t}$, and $u = \frac{dx}{dt}$, also by substitution of (9.9) in (9.8) the following equation is extracted,

$$\frac{\partial}{\partial t}\frac{\partial x_i}{\partial t} = -\frac{(1-\alpha)}{\rho AL}\frac{\partial^2 x_i}{\partial t^2} + \frac{64\tilde{\upsilon}}{2gD^2}\frac{\partial x_i}{\partial t} + F_b \tag{9.10}$$

Therefore,

$$-\frac{(1-\alpha)+\rho AL}{\rho AL}\frac{\partial^2 x_i}{\partial t^2}+\frac{64\tilde{\upsilon}}{2gD^2}\frac{\partial x_i}{\partial t}+F_b=0 \tag{9.11}$$

If an external force, f_p generates by pump, (9.11) can be rewritten as follows,

$$\Theta\ddot{x}+\Upsilon\dot{x}+f_b=f_p \tag{9.12}$$

where $\mathrm{x}\in\mathbb{R}^2$, $\Theta\in\mathbb{R}^{2\times2}$, $\Upsilon\in\mathbb{R}^{2\times2}$, $\mathrm{f}_b=[f_{b1}f_{b2}]^T\in\mathbb{R}^{2\times1}$, $f_p=\left[f_{p1}f_{p2}\right]^T\in\mathbb{R}^{2\times1}$.

Since in this work two pipelines are used, (9.12) can be rewritten as,

$$\begin{aligned}\gamma_1\ddot{x}_1+\varphi_1\dot{x}+f_{b1}&=f_{p1}\\ \gamma_2\ddot{x}_1+\varphi_2\dot{x}+f_{b2}&=f_{p2}\end{aligned} \tag{9.13}$$

Since the pump is supplying pressure to the pipes for the maintaining the flow rates, so the pipes will have the same external force, $f_{p1}=f_{p2}$.

9.2.2 *Modelling of the Actuator*

In order to reduce the vibrations of the motor caused by the external forces (f_p) a torsional actuator is placed on the motor, see Fig. 9.2.

The motor and the pump are interconnected with the help of a shaft. The main purpose of the motor is to drive the pump. The pump with the help of the motor initiate a flow of fluid in the pipe. Any unwanted vibration in the motor will result in the vibration in the pump, which will result in improper flows in the pipe. Therefore, it is important to control the vibration in the motor, so as to control the vibration in the pump for making a stable flow of fluid in the pipelines. For this purpose, a torsional actuator having a motor and disk arrangement as shown in Fig. 9.2 is placed

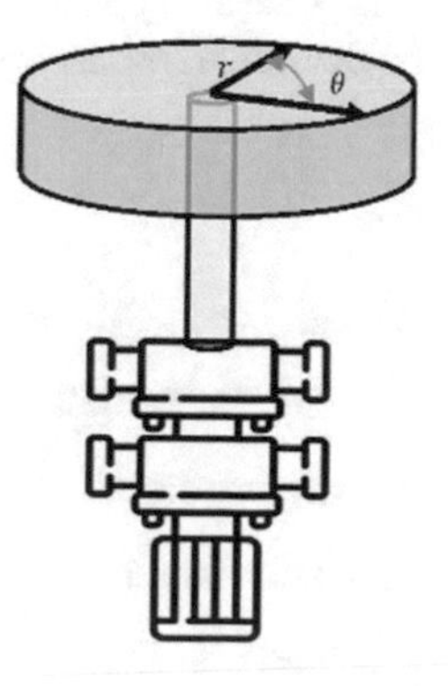

Fig. 9.2 Torsional actuator with motor-pump arrangement

on the top base of the pump The main intention of the torsional actuator is to control the vibration on motor and consequently controlling the flow rates in pipelines.

The inertia moment of the torsional actuator is defined as,

$$J_t = m_t r_t^2 \tag{9.14}$$

where m_t is considered to be the mass of the disc, and r_t is the radius of the disc. The torque produced by means of the disc is defined as

$$u_\alpha = J_t(\ddot{\alpha}_t + \ddot{\alpha}) \tag{9.15}$$

where $\ddot{\alpha}$ is taken to be the angular acceleration of the motor and $\ddot{\alpha}_t$ is taken to be the angular acceleration of the torsional actuator.

In order to reduce the torsional response, the directions of $\ddot{\alpha}_t$ as well as $\ddot{\alpha}$ are taken to be different. The friction of the torsional actuator is defined as [59],

$$f_d = c\dot{\alpha}_t + F_c \tanh(\beta\dot{\alpha}_t) \tag{9.16}$$

where c is taken to be the torsional viscous friction coefficient, β is the motor constant, F_c is taken to be the coulomb friction torque, also tanh is considered to be the hyperbolic tangent which depends on β and motor speed. The final torsion control is expressed as,

$$u_\alpha = J_t(\ddot{\alpha}_t + \ddot{\alpha}) - f_d \tag{9.17}$$

9.2.3 Modelling of the Pump

The general equation of the pump supplying pressure to the pipe for flow control can be demonstrated as [59],

$$T\dot{\omega} = \tau - \left(\tau_p - n\omega\right) \tag{9.18}$$

where,

ω is angular velocity,

τ is motor torque,

τ_p is frictional torque of the motor,

n is load constant,

T is rotations inertia time constant.

Equation (9.18) can be modified as follows,

$$\ddot{x}_p = \frac{\tau - (\tau_p - nx_p)}{T} \tag{9.19}$$

where $\ddot{x}_p$ is the flow acceleration of the pump. Since τ, T, τ_p, and n are known quantities of pump so $\ddot{x}_p$ can be estimated.

The external force generated by the pump is

$$f_p = m_p \ddot{x}_p \tag{9.20}$$

where $\ddot{x}_p$ is the acceleration of the motor and m_p is volumetric mass of the pump.

The shape force vector f_b can be modeled as a linear or a nonlinear model.

From (9.12), by considering shape force vector f_b as non-linear model, the following analysis is illustrated:

In simple non-linear case, (9.12) becomes

$$\Theta \ddot{X} + \Upsilon \dot{X} + f_b = f_p \tag{9.21}$$

where f_b is taken to be non-linear.

9.3 The Tuning Method Based on PD and PID Controller

Since 1700s the control of continuous process has been carried out by utilizing feedback loop. System with feedback control contains drawback which is related to the instability of the system. In order to resolve this problem an appropriate controller should be chosen and also it must be ideal for the monitoring system. The proportional feedback control is uncomplicated and relatively easy to implement. Nevertheless, its outcome is completely sensitive to the sensing location as well as feedback gain. It is concluded from the control theory that these drawbacks of the P control should be overcome by adopting *I* as well as *D* controls. Nevertheless, there exist very few studies investigating the application of the PID control for fluid-mechanics problems. In addition, there are a very limited number of studies dealing with the *P* control and PI controller for pipeline. Due to this lack of investigations, this chapter aims to develop a PID control for flow rate in the pipeline.

The control mechanism is demonstrated in Fig. 9.3 which shows the entire control process of the flow rate of the heavy-oil in pipelines.

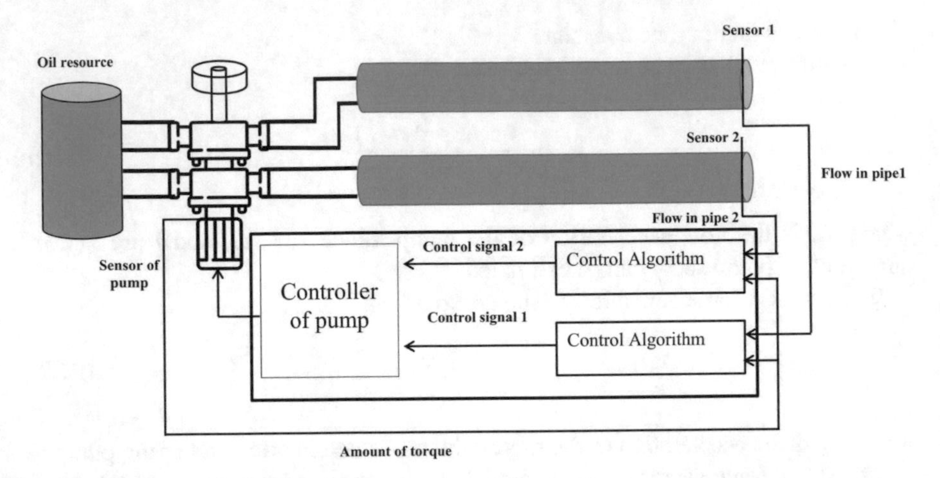

Fig. 9.3 Structure of system

9.3.1 PD Controller

The proportional feedback is a simplified control and easy to implement. Other advantageous of using P controller can be mentioned as minimized dead time, rise time, peak error and integrated error. Furthermore, its outcome is notably sensitive for the sensing location and feedback gain. The drawbacks of P control are:

- Abnormal variations in output upset operators
- Abnormal variations in output upset other loops
- Amplification of noise.

A PD controller consists of proportional element and the derivative element. The PD control is a control law on the basis of only that the output is available for feedback, and is usually common and practical control methodology utilized in the control society.

In the PD control (Fig. 9.4), the controller is composed of a simple gains as illustrated below [60–64]:

$$\Psi(t) = -k_p e(t) - k_d \dot{e}(t) \tag{9.22}$$

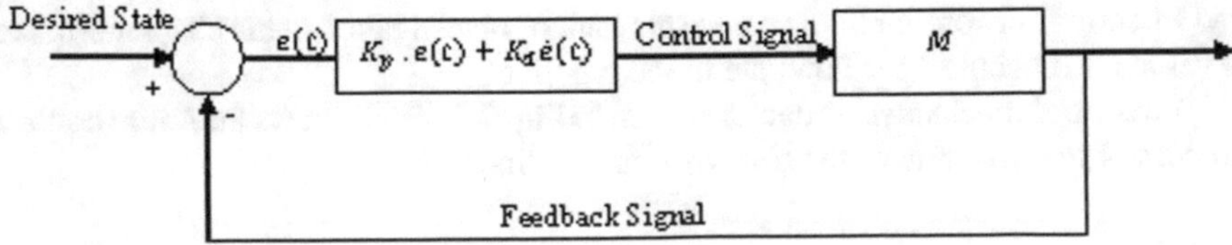

Fig. 9.4 PD controller

where Ψ is the control input, $e(t)$ is the error, k_p is the proportional gain and k_d is the differential gain.

The proportional part $k_p(e(t))$ in the PD control tunes the output signal in straight proportion to the controller input which is the error signal. In a case that there is an increment in the proportional gain k_p, the system under observation results in a fast response, minute steady-state errors as well as an extremely oscillatory response. The differential part $(k_d\dot{e}(t))$ utilizes the amount of variation of the error signal. It presents an element of prediction into the control action, and is capable of making the error to diminish to zero without oscillations with extreme amplitudes. The PD control is designed based on the choosing a proper gains k_p and k_d, mentioned in (9.17), in such a manner that the closed-loop system be stable, also superior performances can be obtained.

The closed-loop system mentioned in (9.12) along with the PD control demonstrated in (9.22) is defined as,

$$\Theta\ddot{X} + \Upsilon\dot{X} + f_b = f_p - k_p e(t) - k_d\dot{e}(t) \tag{9.23}$$

In this case, $e(t) = x - x^d$, and x^d is desired reference. For the flow control $x^d = 0$. k_p as well as k_d are positive constant which correspond to the proportional and derivative gains respectively.

In (9.23) the term $\left(f_b - f_p\right)$ can be considered as uncertainties, and is assumed to satisfy in the Lipschitz condition.

The closed-loop system mentioned in (9.23) can be demonstrated with PD control as,

$$\Theta\ddot{X} + \Upsilon\dot{X} + F_g = -k_p e(t) - k_d\dot{e}(t) \tag{9.24}$$

where $F_g = f_b - f_p - f_d$.

In order to decrease the regulation error, the derivative gain should be increased. The utilization of large derivative gain results in slow temporary performance. It is notable that, if the derivative gain approaches towards infinity, the regulation error approaches to zero [65]. So it is advised to utilize a smaller derivative gain if the system is embedded with high-frequency noise signals.

Theorem *By taking into account the flow system defined in (9.12) and controlled utilizing the PD controller (9.24), the closed-loop system stated in (9.23) is stable under the consideration that the control gains are positive. So the regulation errors approach to the below mentioned residual,*

$$\dot{D} = \left[\dot{X}, \left\|\dot{X}\right\|^2 \le \mu\right] \tag{9.25}$$

where $F_g^T\Lambda^{-1}F_g \le \mu$ and $\Upsilon - \Lambda > 0$, Λ is any positive matrix.

Proof Using (9.12) and (9.22) the new closed loop system is

$$\Theta\ddot{X} + \Upsilon\dot{X} + F_g = -k_p X - k_d \dot{X} \tag{9.26}$$

For the purpose of stability analysis, a Lyapunov candidate is selected as stated in (9.27). The first term of equation denotes the kinetic energy and also the second term signifies the elastic potential energy. Since Θ as well ad k_p are positive definite, therefore $V \geq 0$.

$$V = \frac{1}{2}\dot{X}^T \Theta \dot{X} + \frac{1}{2}\dot{X}^T k_p X \tag{9.27}$$

The derivative of (9.27) is defined as,

$$\begin{aligned} \dot{V} &= \dot{X}^T \Theta \ddot{X} + \dot{X}^T k_p X = \dot{X}^T \left(-\Upsilon \dot{X} - F_g - k_p X - k_d \dot{X}\right) + \dot{X}^T k_p X \\ &= -\dot{X}^T (\Upsilon + k_d)\dot{X} - \dot{X}^T F_g \end{aligned} \tag{9.28}$$

For matrix the inequality of below equation is validated [66],

$$A^T B + B^T A \leq A^T \Lambda A + B^T \Lambda^{-1} B \tag{9.29}$$

The inequality of (9.29) is valid for any $A, B \in \mathbb{R}^{n\times m}$ and any $0 < \Lambda = \Lambda^T \in \mathbb{R}^{n\times n}$, hence the scalar variable $\dot{X}^T F_g$ can be stated as,

$$\dot{X}^T F_g = \frac{1}{2}\dot{X}^T F_g + \frac{1}{2}F_g^T \dot{X} \leq \dot{X}^T \Lambda \dot{X} + F_g^T \Lambda^{-1} F_g \tag{9.30}$$

where Λ is stated as,

$$\Upsilon > \Lambda > 0 \tag{9.31}$$

By applying (9.31) in (9.30), the following relation is extracted,

$$\dot{V} \leq -\dot{X}^T (\Upsilon + k_d - \Lambda)\dot{X} + F_g^T \Lambda^{-1} F_g \tag{9.32}$$

If $k_d > 0$ and $\Upsilon > 0$, so,

$$\dot{V} \leq -\dot{X}^T \Pi \dot{X} + \mu \leq -\lambda\Pi \left\|\dot{X}\right\|^2 + F_g^T \Lambda^{-1} F_g \tag{9.33}$$

where $\Pi = \Upsilon + k_d - \Lambda > 0$.

The boundedness of $F_g^T \Lambda^{-1} F_g \leq \mu$ signifies that the regulation error $\left\|\dot{X}\right\|$ is bounded [67], so,

$$\left\|\dot{X}\right\|^2 > \mu \quad \forall t \in [0, T] \tag{9.34}$$

Therefore it can be concluded that $\dot{V} < 0$ when $\|\dot{X}\|^2 > \mu$. The next step is to demonstrate that the total time during which $\|\dot{X}\|^2 > \mu$ is finite.

Suppose T signifies the time interval during which $\|\dot{X}\|^2 > \mu$. $\|\dot{X}\|^2 > \mu$ is inside the circle if $\|\dot{X}\|^2 > \mu$ is outside the circle of radius μ for finite time and afterward re-enters the circle. Furthermore, as the total time $\|\dot{X}\|^2 > \mu$ is finite, so $\sum_{k=1}^{\infty} T_k < \infty$, also,

$$\lim_{k\to\infty} T_k = 0 \tag{9.35}$$

Therefore, $\|\dot{X}\|^2$ is bounded with an invariant set argument. Furthermore, by utilizing (9.33) it can be proved that $\|\dot{X}\|$ is likewise bounded. Assume $\|\dot{X}\|^2$ to be the greatest tracking error throughout the T_k interval. Therefore, by utilizing (9.28) and since $\|\dot{X}\|^2$ is bounded so we have,

$$\lim_{k\to\infty}\left[\|\dot{X}\|^2 - \mu\right] = 0 \tag{9.36}$$

Hence, $\|\dot{x}\|^2$ approaches to μ_{f_b}, thus, the derivative of regulation error x approaches to the residual set,

$$\dot{D} = \left[\dot{X}, \|\dot{X}\|^2 \leq \mu\right] \tag{9.37}$$

Besides, for $\|\dot{X}\|^2 > \mu$ the total time is finite, therefore $V = \frac{1}{2}\dot{X}^T\Theta\dot{X} + \frac{1}{2}\dot{X}^T k_p X$ is bounded, thus the regulation error $\dot{X}$ is bounded.

9.3.2 PID Controller

The PID control is considered as a control law in which the existence of output for feedback is essential. This practical control method is widely utilized in the control society. In the PID control, the controller is made of a simple gain (P control), an integrator (I control), a differentiator (D control) or some weighted composition of these possibilities [28], see Fig. 9.5.

The PID control is expressed as,

$$\Psi(t) = -\kappa_p e(t) - \kappa_i \int_0^t e(t)d\tau - \kappa_d \dot{e}(t) \tag{9.38}$$

Fig. 9.5 PID controller

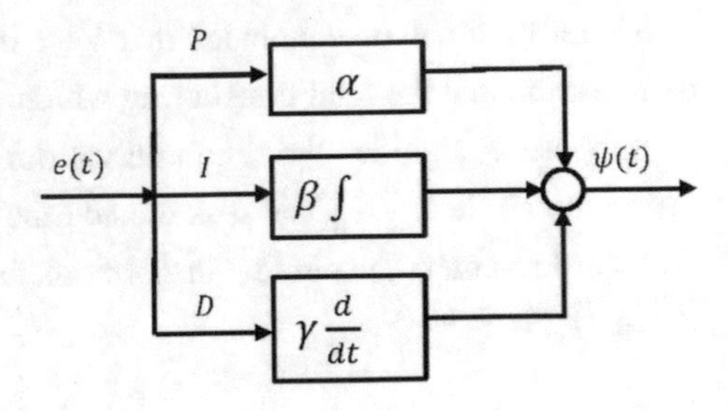

where k_p, k_i, as well as k_d are positive definite and k_i is the integration gain. For the flow control, X^d is desired reference and also $X^d = \dot{X}^d = 0$. Hence, Eq. (9.38) is rewritten as below,

$$\Psi(t) = -\kappa_p X - \kappa_i \int_0^t X d\tau - \kappa_d \dot{X} \tag{9.39}$$

For analyzing the PID controller, Eq. (9.38) can be stated as below,

$$\begin{aligned} \Psi(t) &= -\kappa_p X - \kappa_d \dot{X} - \vartheta \\ \vartheta &= \kappa_i \int_0^t X d\tau, \quad \vartheta(0) = 0 \end{aligned} \tag{9.40}$$

The closed-loop system of Eq. (9.12) along with the PID control (Eq. 9.39) is demonstrated as below,

$$\begin{aligned} \Theta\ddot{x} + \Upsilon\dot{x} + F_g &= -\kappa_p X - \kappa_d \dot{X} - \vartheta \\ \dot{\vartheta} &= \kappa_i X \end{aligned} \tag{9.41}$$

In matrix form, the closed-loop system is defined as,

$$\frac{d}{dt}\begin{bmatrix} \vartheta \\ X \\ \dot{X} \end{bmatrix} = \begin{bmatrix} \kappa_i X \\ \dot{X} \\ -\Theta^{-1}\left(\Upsilon\dot{X} + F_g + k_p X + k_d \dot{X} + \vartheta\right) \end{bmatrix} \tag{9.42}$$

Here the stability of the PID control demonstrated by Eq. (9.39) is analyzed. The equilibrium of Eq. (9.42) is presented by $\left[\vartheta \ X \ \dot{X}\right] = \left[\hat{\vartheta} \ 0 \ 0\right]$. As at equilibrium point $X = 0$ as well as $\dot{X} = 0$, the equilibrium is $\left[f(0),\ 0,\ 0\right]$. For moving the equilibrium to the origin, the following is defined,

$$\hat{\vartheta} = \vartheta - f(0) \tag{9.43}$$

Therefor, the final closed-loop equation is defined as,

$$\begin{aligned} &\Theta\ddot{X} + \Upsilon\dot{X} + F_g = -\kappa_p X - \kappa_d \dot{X} - \vartheta + f(0) \\ &\vartheta = \kappa_i x \end{aligned} \tag{9.44}$$

For analyzing the stability of Eq. (9.28), the following properties are required,

Property 1 *The positive definite matrix* Θ *should satisfy in the below condition*

$$0 < \lambda_{min}(\Theta) \leq \|\Theta\| \leq \lambda_{Max}(\Theta) \leq \overline{\gamma} \tag{9.45}$$

such that $\lambda_{min}(\Theta)$ *as well as* $\lambda_{Max}(\Theta)$ *are considered as the minimum and maximum eigenvalues of the matrix* Θ*, respectively also* $\overline{\gamma} > 0$ *is taken to be the upper bound.*

Property 2 *f is taken to be Lipschitz over* $\tilde{x}$ *and* $\tilde{y}$ *if*

$$\|f(\tilde{x}) - f(\tilde{y})\| \leq \Omega\|\tilde{x} - \tilde{y}\| \tag{9.46}$$

As F_g is first-order continuous functions also satisfies in Lipschitz condition, Property 2 is hereby established.

The lower bound of F_g can be calculated as below,

$$\int_0^t f dx = \int_0^t F_g dx + \int_0^t d_u dx \tag{9.47}$$

The lower bound of $\int_0^t F_g dx$ is stated as $-\hat{F}_g$ also $\int_0^t d_u dx$ as $-\hat{D}_u$. Therefore, the lower bound of Ω is defined as,

$$\Omega = -\hat{F}_g - \hat{D}_u \tag{9.48}$$

The stability analysis of PID control approach is given by the below mentioned theorem.

Theorem *By taking into consideration the structural system of Eq.* (9.12) *controlled by the PID control approach of Eq.* (9.38)*, the closed-loop system of Eq.* (9.44) *is taken to be asymptotically stable at the equilibriums* $\left[\vartheta - f(0), X, \dot{X}\right]^T = 0$*, if the following gains are satisfied,*

$$\begin{aligned} \lambda_{min}(\kappa_d) \geq \frac{1}{4}\left(\frac{1}{3}\lambda_{min}(\Theta)\lambda_{min}(\kappa_p)\right)^{1/2}\left[1 + \frac{k_e}{\lambda_{Max}(\Theta)}\right] \\ - \lambda_{min}(\Upsilon) \\ \lambda_{Max}(\kappa_i) \leq \frac{1}{6}\left(\frac{1}{3}\lambda_{min}(\Theta)\lambda_{min}(\kappa_p)\right)^{1/2}\left[\frac{\lambda_{min}(\kappa_p)}{\lambda_{Max}(\Theta)}\right] \end{aligned}$$

$$\lambda_{min}(\kappa_p) \geq \frac{3}{2}[\Omega + \Xi] \tag{9.49}$$

Proof The Lyapunov function can be stated as below,

$$V = \frac{1}{2}\dot{X}^T \Theta \dot{X} + \frac{1}{2}X^T \kappa_p X + \frac{\sigma}{4}\vartheta^T \kappa_i^{-1}\vartheta + X^T\vartheta$$
$$+ \frac{\sigma}{2}X^T \kappa_d X + \frac{\sigma}{4}X^T \kappa_d X + \int_0^t F_g dx - \Omega \tag{9.50}$$

where $V(0) = 0$. For representing that $V \geq 0$, we divide it into three parts, $V = V_1 + V_2 + V_3$, where,

$$V_1 = \frac{1}{6}X^T \kappa_p X + \frac{\sigma}{4}X^T \kappa_d X + \int_0^t F_g dx - \Omega \geq 0, \quad \kappa_p > 0,\ \kappa_d > 0 \tag{9.51}$$

$$V_2 = \frac{1}{6}X^T \kappa_p X + \frac{\sigma}{4}\vartheta^T \kappa_i^{-1}\vartheta + X^T\vartheta$$
$$\geq \frac{1}{2}\frac{1}{3}\lambda_m(\kappa_p)\|x\|^2 + \frac{\sigma\lambda_{min}(\kappa_i^{-1})}{4}\|\vartheta\|^2$$
$$- \|x\|\|\vartheta\| \tag{9.52}$$

In a case that $\sigma \geq \frac{3}{(\lambda_{min}(\kappa_i^{-1})\lambda_{min}(\kappa_p))}$, the following is obtained,

$$V_2 \geq \frac{1}{2}\left(\sqrt{\frac{\lambda_{min}(\kappa_p)}{3}}\|x\| - \sqrt{\frac{3}{4(\lambda_{min}(\kappa_p))}}\|\xi\|\right)^2 \geq 0 \tag{9.53}$$

Also,

$$V_3 \geq \frac{1}{6}X^T k_p X + \frac{1}{2}\dot{X}^T \Theta \dot{X} + \frac{\sigma}{2}X^T \Theta \dot{X} \tag{9.54}$$

Since

$$X^T AX \geq \|X\|\|AX\| \geq \|X\|\|A\|\|X\| \geq \lambda_{Max}(A)\|X\|^2 \tag{9.55}$$

in a case that $\sigma \leq \frac{1}{2}\frac{\sqrt{\frac{1}{3}\lambda_{min}(\Theta)\lambda_m(\kappa_p)}}{\lambda_{Max}(\Upsilon)}$, the following is obtained,

$$V_3 \geq \frac{1}{2}\begin{pmatrix} \frac{1}{3}\lambda_{min}(\kappa_p)\|X\|^2 + \lambda_{min}(\Theta)\|\dot{X}\|^2 \\ +\sigma\lambda_{Max}(\Theta)\|X\|\|\dot{X}\| \end{pmatrix}$$

$$= \frac{1}{2}\left(\sqrt{\frac{\lambda_{min}(\kappa_p)}{3}}\|X\| + \sqrt{\lambda_{Max}(\Theta)}\|\dot{X}\|\right)^2 \geq 0 \tag{9.56}$$

Therefore,

$$\frac{1}{2}\frac{\sqrt{\frac{1}{3}\lambda_{min}(\Theta)\lambda_m(\kappa_p)}}{\lambda_{Max}(\Theta)} \geq \sigma \geq \frac{3}{\left(\lambda_{min}(\kappa_i^{-1})\lambda_{min}(\kappa_p)\right)} \tag{9.57}$$

The derivative of Eq. (9.40) is obtained as below,

$$\begin{aligned}\dot{V} = {} & \dot{X}^T\Theta\ddot{X} + \dot{X}^T\kappa_p X + \frac{\sigma}{2}\vartheta^T\kappa_i^{-1}\vartheta + \dot{X}^T\vartheta + \dot{X}^T\vartheta + X^T\vartheta \\ & + \frac{\sigma}{2}\dot{X}^T\Theta\dot{X} + \frac{\sigma}{2}X^T\Theta\ddot{X} + \sigma\dot{X}^T\kappa_d X \\ & + \dot{X}^T F_g\end{aligned} \tag{9.58}$$

For matrix the inequality of below equation is validated {Yang, 2001 #765},

$$A^T B + B^T A \leq A^T\Lambda A + B^T\Lambda^{-1}B \tag{9.59}$$

The inequality of (9.58) is valid for any $A,\ B \in \mathbb{R}^{n\times m}$ and any $0 < \Lambda = \Lambda^T \in \mathbb{R}^{n\times n}$, hence the scalar variable $\dot{X}^T F_g$ can be stated as,

$$\dot{X}^T F_g = \frac{1}{2}\dot{X}^T F_g + \frac{1}{2}F_g^T\dot{X} \leq \dot{X}^T\Lambda_{F_g}\dot{X} + F_g^T\Lambda_{F_g}^{-1}F_g \tag{9.60}$$

Utilizing Eq. (9.58) the following is obtained,

$$-\frac{\sigma}{2}X^T\Upsilon\dot{X} \leq \frac{\sigma}{2}\Xi_\Upsilon\left(X^T x + \dot{X}^T\dot{X}\right) \tag{9.61}$$

where $\|\Upsilon\| \leq \Xi_\Upsilon$. Therefore, $\vartheta = k_i,\ \vartheta^T k_i^{-1}\vartheta$ becomes $x^T\vartheta$ also $x^T\vartheta$ becomes $x^T k_i$. Utilizing Eq. (9.61) the following is extracted,

$$\begin{aligned}\dot{V} = {} & -\dot{X}^T\left[\Upsilon + \kappa_d - \frac{\sigma}{2}\Theta - \frac{\sigma}{2}\Xi\right]\dot{X} \\ & - X^T\left[\frac{\sigma}{2}\kappa_p - \kappa_i - \frac{\sigma}{2}\Xi\right]X \\ & - \frac{\sigma}{2}X^T\left[F_g - f(0)\right] + \dot{X}^T f(0)\end{aligned} \tag{9.62}$$

By applying the Lipschitz condition of Eq. (9.46) the following is obtained,

$$\frac{\sigma}{2}X^T\left[f(0) - F_g\right] \leq \frac{\sigma}{2}\kappa_f\|X\|^2 \tag{9.63}$$

$$-\frac{\sigma}{2}X^T\left[F_g - f(0)\right] \leq X^T\frac{\sigma}{2}\Omega X \tag{9.64}$$

From Eq. (9.58) we have,

$$\dot{X}^T f(0) \geq -f^T(0)\Lambda^{-1}f(0) \tag{9.65}$$

By utilizing Eq. (9.54)

$$\begin{aligned}\dot{V} = &-\dot{X}^T\left[\Upsilon + \kappa_d - \frac{\sigma}{2}\Theta - \frac{\sigma}{2}\Xi\right]\dot{X}\\ &- X^T\left[\frac{\sigma}{2}\kappa_p - \kappa_i - \frac{\sigma}{2}\Xi - \frac{\sigma}{2}\kappa_f\right]X\\ &+ \dot{X}^T f(0)\end{aligned} \tag{9.66}$$

Becomes

$$\begin{aligned}\dot{V} \leq &-\dot{X}^T\left[\lambda_{min}(\Upsilon) + \lambda_{min}(\kappa_d) - \frac{\sigma}{2}\lambda_{Max}(\Theta) - \frac{\sigma}{2}\Xi\right]\dot{X}\\ &- X^T\left[\frac{\sigma}{2}\lambda_{min}\left(\kappa_p\right) - \lambda_{min}(\kappa_i) - \frac{\sigma}{2}\Xi - \frac{\sigma}{2}\Omega\right]X\end{aligned} \tag{9.67}$$

Therefore, $\dot{V} \leq 0$, $\|X\|$ diminishes if the following conditions are held:

$$\lambda_{min}(\Upsilon) + \lambda_{min}(\kappa_d) \geq \frac{\sigma}{2}[\lambda_{Max}(\Theta) + \kappa_c]$$

$$\lambda_{min}\left(k_p\right) \geq \frac{2}{\sigma}\lambda_{Max}(k_i) + \Xi + \Omega$$

By utilizing Eq. (9.57) as well as $\lambda_{min}\left(\kappa_i^{-1}\right) = \frac{1}{\lambda_{Max}(\kappa_i)}$, the following is obtained,

$$\begin{aligned}\lambda_{min}(\kappa_d) \geq &\frac{1}{4}\left(\frac{1}{3}\lambda_{min}(\Theta)\lambda_m\left(\kappa_p\right)\right)^{1/2}\left[1 + \frac{\Xi}{\lambda_{Max}(\Theta)}\right]\\ &- \lambda_{min}(\Upsilon)\end{aligned} \tag{9.68}$$

Again $\frac{2}{\sigma}\lambda_{Max}(k_i) = \frac{2}{3}\lambda_{min}\left(k_p\right)$. Thus,

$$\lambda_{Max}(\kappa_i) \leq \frac{1}{6}\left(\frac{1}{3}\lambda_{min}(\Theta)\lambda_{min}\left(\kappa_p\right)\right)^{1/2}\frac{\lambda_{min}\left(\kappa_p\right)}{\lambda_{Max}(\Theta)} \tag{9.69}$$

Furthermore,

$$\lambda_{min}\left(\kappa_p\right) \geq \frac{3}{2}[\Omega + \Xi] \tag{9.70}$$

9.4 Numerical Analysis

For the numerical analysis purpose and for the validation of the novel control strategy, the various parameters associated with the flow control are illustrated as Table 9.1.

The implemented software in this chapter is Matlab/Simulink. Simulations are presented to show that the motor vibration can be attenuated to a significant level by using the torsional actuator with the developed controllers thus validating the effectiveness of the proposed control approach using PD and PID controllers. A simulation period of 20 s was considered for evaluation. For the simulation purposes, the weight of the torsional actuator is considered to be 5% of the motor and pump weight in combination.

The Theorem proposed in this chapter generates sufficient conditions for the minimum amounts of the proportional as well as the derivative gains. This Theorem validates that both proportional and derivative gain must be positive as negative gains can make the systems unstable. The PD gains are selected in order to ensure the efficiency and within the stable range by the stability analysis.

Since the maximum flow rate of the pipeline is 13 m^3/s so the other non linear force associated with Ω has to be less than 13 m^3/s. Hence, we select the value of Ω = 13 m^3/s, and

$$\lambda_{min}(\gamma_1) = 1.0002,\ \lambda_{min}(\gamma_2) = 1.000206,\ \lambda_{min}(\Xi_1) = 2.09,\ \lambda_{min}(\Xi_2) = 2.09$$

For pipe 1 (see Fig. 9.3) the gain values are present as

$$\lambda_{min}(k_{p1}) \geq 453,\ \lambda_{min}(k_{d1}) \geq 8. \tag{9.71}$$

Also for pipe 2 (see Fig. 9.3) the gain values are present as

$$\lambda_{min}(k_{p2}) \geq 453,\ \lambda_{min}(k_{d2}) \geq 8. \tag{9.72}$$

Table 9.1 Parameters associated with the flow control

Nomenclature		
ρ	1240	kg/m^3
υ	1.604×10^{-3}	m^2/s
g	9.81	m/s^2
L	100	m
α	0.95	–
D	0.05	m
A	1.96×10^{-3}	m^2
r_t	0.3	m
m_t	1.5	kg

From ranges mentioned in (9.71) and (9.72), and with several trials for PD controller the best value of gains are selected as,

$$k_{p1} = 510,\ k_{d1} = 73,\ k_{p2} = 520,\ k_{d2} = 83$$

From Theorem 2, we use the following PID gains

$$\lambda_{min}(\kappa_{p1}) \geq 453,\ \lambda_{min}(\kappa_{d1}) \geq 8,\ \lambda_{Max}(\kappa_{i1}) \leq 928 \tag{9.73}$$

Also for pipe 2 we have,

$$\lambda_{min}(\kappa_{p2}) \geq 453,\ \lambda_{min}(\kappa_{d2}) \geq 8,\ \lambda_{Max}(\kappa_{i2}) \leq 928 \tag{9.74}$$

From ranges mentioned by Eqs. (9.73) and (9.74), the best value of gains are

$$\kappa_{p1} = 458,\ \kappa_{d1} = 110,\ \kappa_{i1} = 650,\ \kappa_{p2} = 480,$$
$$\kappa_{d2} = 115,\ \kappa_{i2} = 645$$

Two subsystem blocks of milling model, one without control system and other with control system are created in order to compare the results. The flow rate from the pump is the input to the flow model. Numerical integrators are used to compute the velocity and position from the acceleration signal. The control signal from the controller subsystem block is fed to the Torsional actuator simulation block to generate the required control forces.

Figures 9.6 and 9.7 represents the vibration attenuate in motor. As can be concluded from these figures that the PD controller has high performance in minimizing vibration. Figures 9.8 and 9.9 represents the flow rate in pipeline 1 and 2, respectively. These Figures demonstrate that with PD controller, the flow rates initiate from zero and maintaining stable flow rate which prove the effectiveness of PD controller.

Figures 9.10 and 9.11 represents the vibration attenuate in motor. From these figures, it can be concluded that PID controller is performing good in minimizing the

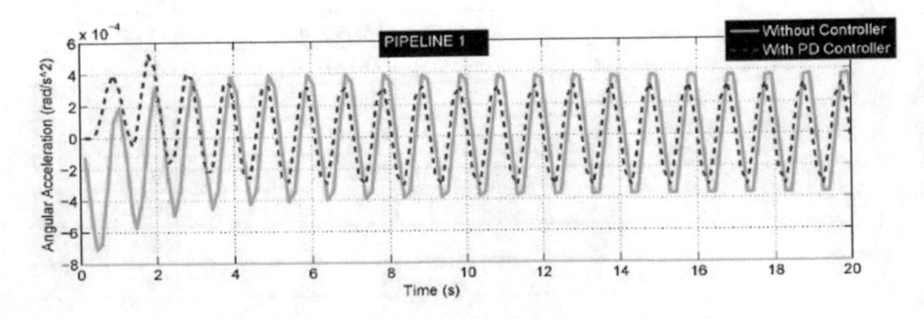

Fig. 9.6 Comparison of motor vibration attenuation using PD controller for pipeline 1

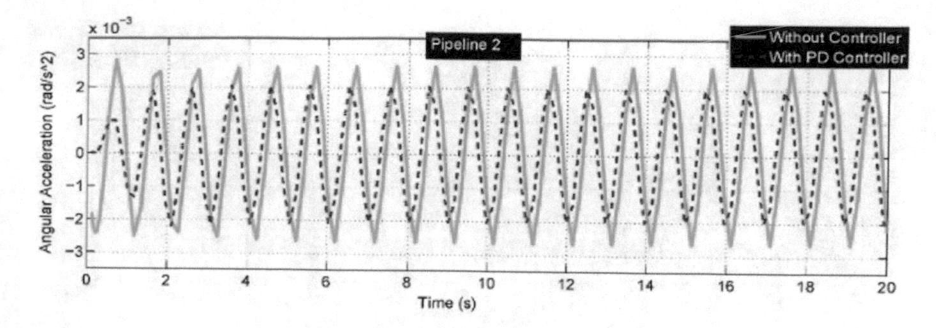

Fig. 9.7 Comparison of motor vibration attenuation using PD controller for pipeline 2

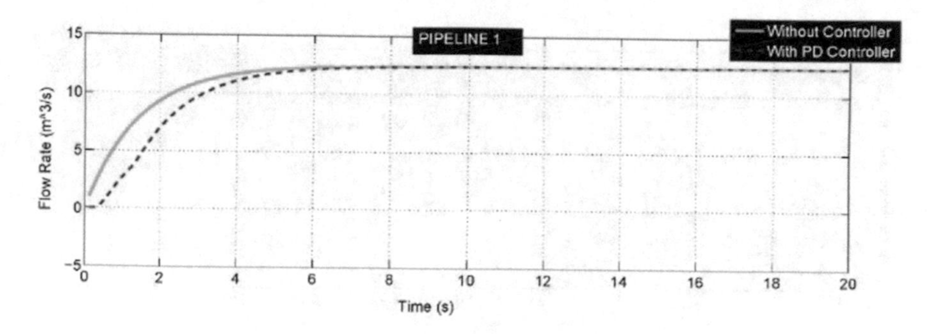

Fig. 9.8 Stability of flow rate with using of controller in pipeline 1

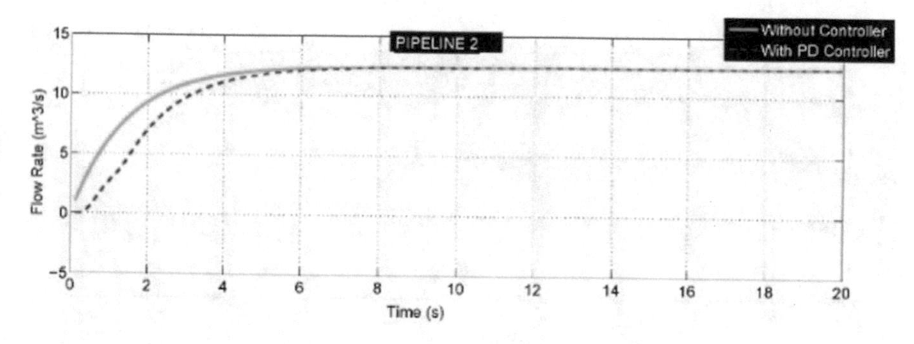

Fig. 9.9 Stability of flow rate with using of controller in pipeline 2

vibration. Figures 9.12 and 9.13 represents the flow rate in pipeline 1 and pipeline 2. When PID controller are used, the flow rates initiate from zero and maintaining stable flow rate, which proves the effectiveness of PID controller.

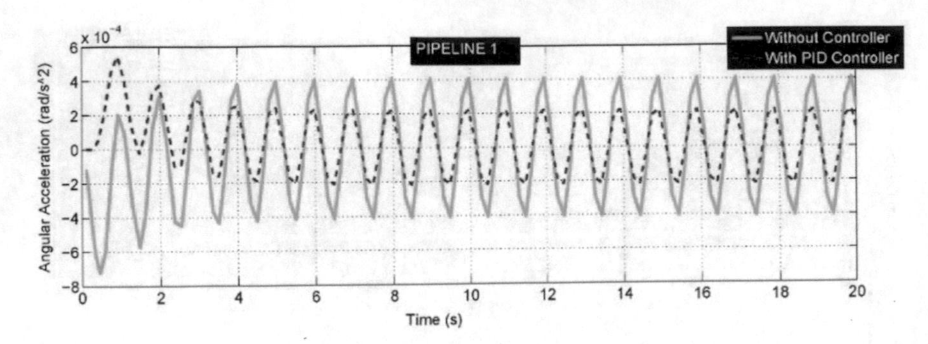

Fig. 9.10 Comparison of motor vibration attenuation using PID controller for pipeline 1

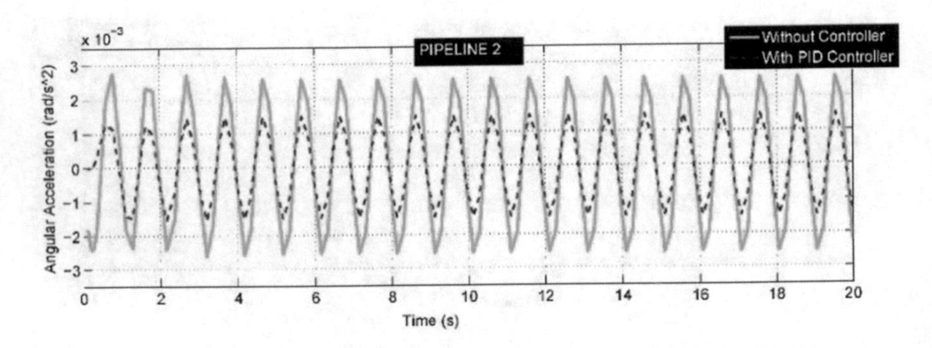

Fig. 9.11 Comparison of motor vibration attenuation using PID controller for pipeline 2

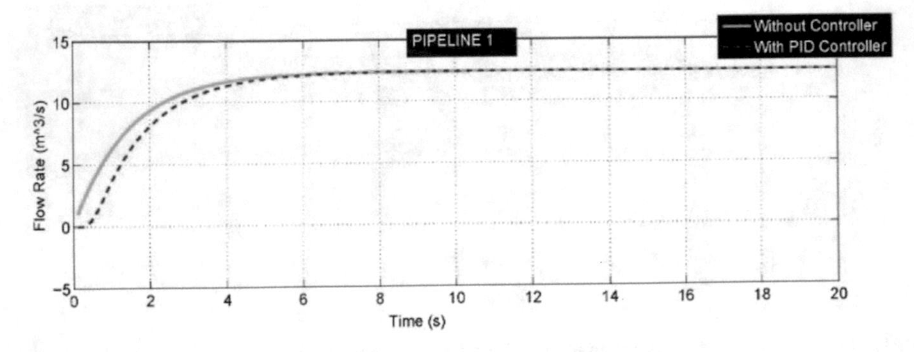

Fig. 9.12 Stability of flow rate with using of PID controller in pipeline 1

9.5 Conclusions

In this chapter, a novel active control strategy for the attenuation of motor vibration is proposed which consequently controls the flow rate in heavy-oil pipelines. The important theoretical contribution associated with the stability analysis for the PD

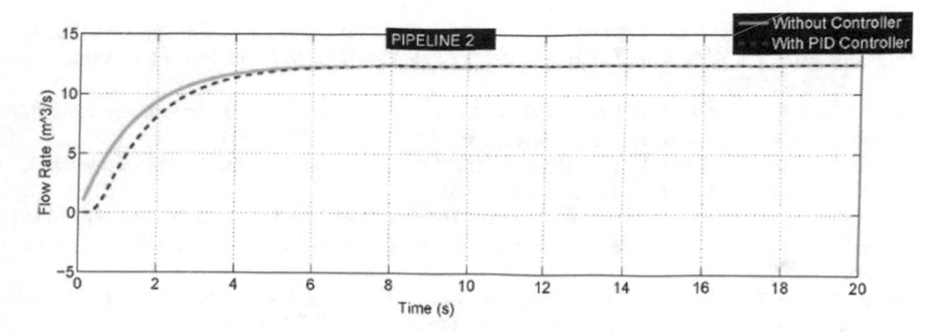

Fig. 9.13 Stability of flow rate with using of PID controller in pipeline 2

and PID controller is developed. The required stability conditions are obtained for the purpose of tuning the PD and PID gains. By utilizing Lyapunov stability analysis, the sufficient conditions for the minimum amounts of the proportional as well as the derivative gains are obtained. The numerical simulation and analysis validates the effectiveness of PD and PID controllers in the minimization of motor vibration to control the flow rate in pipelines.

References

1. Åström, K.J., Wittenmark, B.: Computer-Controlled Systems: Theory and Design. Courier Corporation (2013)
2. Ziegler, J.G., Nichols, N.B.: Optimum settings for automatic controllers. J. Dyn. Syst. Meas. Control **115**(2B), 220–222 (1993). https://doi.org/10.1115/1.2899060
3. Basilio, J.C., Matos, S.: Design of PI and PID controllers with transient performance specification. IEEE Trans. Educ. **45**(4), 364–370 (2002)
4. Rivera, D.E., Morari, M., Skogestad, S.: Internal model control: PID controller design. Ind. Eng. Chem. Process Des. Dev. **25**(1), 252–265 (1986)
5. Basilio, J., Silva, J., Rolim, L., Moreira, M.: H∞ design of rotor flux-oriented current-controlled induction motor drives: speed control, noise attenuation and stability robustness. IET Control Theory Appl. **4**(11), 2491–2505 (2010)
6. Suh, I., Bien, Z.: Proportional minus delay controller. IEEE Trans. Autom. Control. **24**(2), 370–372 (1979)
7. Berger, E.: Suppression of vortex shedding and turbulence behind oscillating cylinders. Phys. Fluids **10**(9), S191–S193 (1967)
8. Baz, A., Ro, J.: Active control of flow-induced vibrations of a flexible cylinder using direct velocity feedback. J. Sound Vib. **146**(1), 33–45 (1991)
9. Williams, J.F., Zhao, B.: The active control of vortex shedding. J. Fluids Struct. **3**(2), 115–122 (1989)
10. Roussopoulos, K.: Feedback control of vortex shedding at low Reynolds numbers. J. Fluid Mech. **248**, 267–296 (1993)
11. Hiejima, S., Kumao, T., Taniguchi, T.: Feedback control of vortex shedding around a bluff body by velocity excitation. Int. J. Comput. Fluid Dyn. **19**(1), 87–92 (2005)
12. Sung, S.W., Lee, J., Lee, I.-B.: Process Identification and PID Control. Wiley (2009)
13. Kusters, A., Van Ditzhuijzen, G.: MIMO system identification of a slab reheating furnace. In: Proceedings of the Third IEEE Conference on Control Applications, pp. 1557–1563 (1994)

14. O'Dwyer, A.: Handbook of PI and PID Controller Tuning Rules. Imperial College Press (2009)
15. Choi, H., Jeon, W.P., Kim, J.: Control of flow over a bluff body. Annu. Rev. Fluid Mech. **40**, 113–139 (2008)
16. Collis, S.S., Joslin, R.D., Seifert, A., Theofilis, V.: Issues in active flow control: theory, control, simulation, and experiment. Prog. Aerosp. Sci. **40**(4–5), 237–289 (2004)
17. Gad-El-Hak, M., Pollard, A., Bonnet, J.-P.: Flow Control: Fundamentals and Practices, vol. 53. Springer Science & Business Media (2003)
18. Fadlun, E., Verzicco, R., Orlandi, P., Mohd-Yusof, J.: Combined immersed-boundary finite-difference methods for three-dimensional complex flow simulations. J. Comput. Phys. **161**(1), 35–60 (2000)
19. Mohd-Yusof, J.: Combined immersed-boundary/B-spline methods for simulations of flow in complex geometries. Cent. Turb. Res. Annu. Res. Briefs **161**(1), 317–327 (1997)
20. Nandy, A., Mondal, S., Chakraborty, P., Nandi, G.C.: Development of a robust microcontroller based intelligent prosthetic limb. In: International Conference on Contemporary Computing, pp. 452–462. Springer (2012)
21. Yu, W., Jafari, R.: Modeling and Control of Uncertain Nonlinear Systems with Fuzzy Equations and Z-Number. Wiley (2019)
22. Razvarz, S., Jafari, R.: ICA and ANN modeling for photocatalytic removal of pollution in wastewater. Math. Comput. Appl. **22**(3), 38 (2017)
23. Jafari, R., Razvarz, S., Gegov, A.: Neural network approach to solving fuzzy nonlinear equations using Z-numbers. IEEE Trans. Fuzzy Syst. (2019)
24. Razvarz, S., Jafari, R.: Intelligent techniques for photocatalytic removal of pollution in wastewater. J. Electr. Eng. **5**(1), 321–328 (2017)
25. Jafari, R., Razvarz, S., Gegov, A., Paul, S., Keshtkar, S.: Fuzzy Sumudu transform approach to solving fuzzy differential equations with Z-numbers. In: Advanced Fuzzy Logic Approaches in Engineering Science, pp. 18–48. IGI Global (2019)
26. Jafari, R., Razvarz, S., Gegov, A.: A novel technique to solve fully fuzzy nonlinear matrix equations. In: International Conference on Theory and Applications of Fuzzy Systems and Soft Computing, pp. 886–892. Springer (2018)
27. Jafari, R., Razvarz, S., Gegov, A.: Fuzzy differential equations for modeling and control of fuzzy systems. In: International Conference on Theory and Applications of Fuzzy Systems and Soft Computing, pp. 732–740. Springer (2018)
28. Jafari, R., Yu, W., Razvarz, S., Gegov, A.: Numerical methods for solving fuzzy equations: a survey. Fuzzy Sets Syst. (2019)
29. Jafari, R., Razvarz, S., Gegov, A.: A new computational method for solving fully fuzzy nonlinear systems. In: International Conference on Computational Collective Intelligence, pp. 503–512. Springer (2018)
30. Jafari, R., Razvarz, S., Gegov, A., Paul, S.: Modeling and control of uncertain nonlinear systems. In: 2018 International Conference on Intelligent Systems (IS), pp. 168–173. IEEE (2018)
31. Jafari, R., Razvarz, S., Gegov, A.: A novel technique for solving fully fuzzy nonlinear systems based on neural networks. Vietnam J. Comput. Sci. **7**(1), 93–107 (2020)
32. Razvarz, S., Hernández-Rodríguez, F., Jafari, R., Gegov, A.: Foundation of Z-numbers and engineering applications. In: Latin American Symposium on Industrial and Robotic Systems, pp. 15–24. Springer (2019)
33. Jafari, R., Contreras, M.A., Yu, W., Gegov, A.: Applications of fuzzy logic, artificial neural network and neuro-fuzzy in industrial engineering. In: Latin American Symposium on Industrial and Robotic Systems, pp. 9–14. Springer (2019)
34. Jafari, R., Razvarz, S., Gegov, A., Yu, W.: Fuzzy control of uncertain nonlinear systems with numerical techniques: a survey. In: UK Workshop on Computational Intelligence, pp. 3–14. Springer (2019)
35. Jafari, R., Razvarz, S., Yu, W., Gegov, A., Goodwin, M., Adda, M.: Genetic algorithm modeling for photocatalytic elimination of impurity in wastewater. In: Proceedings of SAI Intelligent Systems Conference, pp. 228–236. Springer (2019)

36. Tatchum, M., Gegov, A., Jafari, R., Razvarz, S.: Parallel distributed compensation for voltage controlled active magnetic bearing system using integral fuzzy model. In: 2018 International Conference on Intelligent Systems (IS), pp. 190–198. IEEE (2018)
37. Razvarz, S., Jafari, R., Gegov, A.: Solving partial differential equations with Bernstein neural networks. In: UK Workshop on Computational Intelligence, pp. 57–70. Springer (2018)
38. Jafarian, A., Jafari, R.: New iterative approach for solving fully fuzzy polynomials. Int. J. Fuzzy Math. Syst. **3**(2), 75–83
39. Jafarian, A., Jafari, R.: New method for solving fuzzy polynomials. Adv. Fuzzy Math. **8**(1), 25–33 (2013)
40. Jafarian, A., Jafari, R.: An iterative method for solving fuzzy polynomials by fuzzy neural networks (2012)
41. Jafarian, A., Jafari, R.: Simulation and evaluation of fuzzy polynomials by feed-back neural networks (2012)
42. Jafari, R., Yu, W.: Fuzzy control for uncertainty nonlinear systems with dual fuzzy equations. J. Intell. Fuzzy Syst. **29**(3), 1229–1240 (2015)
43. Jafari, R., Yu, W.: Fuzzy modeling for uncertainty nonlinear systems with fuzzy equations. Math. Probl. Eng. (2017)
44. Müller, S., Milano, M., Koumoutsakos, P.: Application of machine learning algorithms to flow modeling and optimization. Annu. Res. Briefs 169–178 (1999)
45. Aly, A.A.: An artificial neural network flow control of variable displacement piston pump with pressure compensation. Int. J. Control Autom. Syst. **4**(1), 1–7 (2015)
46. Razvarz, S., Vargas-Jarillo, C., Jafari, R.: Pipeline monitoring architecture based on observability and controllability analysis. In: 2019 IEEE International Conference on Mechatronics (ICM), 18–20 Mar 2019, pp. 420–423 (2019)
47. Jafari, R., Razvarz, S., Vargas-Jarillo, C., Yu, W.: Control of flow rate in pipeline using PID controller. In: 2019 IEEE 16th International Conference on Networking, Sensing and Control (ICNSC), 9–11 May 2019, pp. 293–298 (2019)
48. Jafari, R., Razvarz, S., Vargas-Jarillo, C., Gegov, A.: Blockage detection in pipeline based on the extended Kalman filter observer. Electronics **9**(1), 91–107 (2020)
49. Razvarz, S., Jafari, R., Vargas-Jarillo, C.: Modelling and analysis of flow rate and pressure head in pipelines. In: 2019 16th International Conference on Electrical Engineering, Computing Science and Automatic Control (CCE), pp. 1–6. IEEE (2019)
50. Jafari, R., Razvarz, S., Vargas-Jarillo, C., Gegov, A.E.: The effect of baffles on heat transfer. In: ICINCO (2), pp. 607–612 (2019)
51. Razvarz, S., Jafari, R., Vargas-Jarillo, C., Gegov, A., Forooshani, M.: Leakage detection in pipeline based on second order extended Kalman filter observer. IFAC-PapersOnLine **52**(29), 116–121 (2019)
52. Razvarz, S., Vargas-Jarillo, C., Jafari, R., Gegov, A.: Flow control of fluid in pipelines using PID controller. IEEE Access **7**, 25673–25680 (2019)
53. Razvarz, S., Chavez, L.F.G., Vargas-Jarillo, C.: Nanotechnology applications in industry and heat transfer. In: Latin American Symposium on Industrial and Robotic Systems, pp. 1–8. Springer (2019)
54. Park, D., Ladd, D., Hendricks, E.: Feedback control of von Kármán vortex shedding behind a circular cylinder at low Reynolds numbers. Phys. Fluids **6**(7), 2390–2405 (1994)
55. Herrán-González, A., De La Cruz, J., De Andrés-Toro, B., Risco-Martín, J.L.: Modeling and simulation of a gas distribution pipeline network. Appl. Math. Model. **33**(3), 1584–1600 (2009)
56. Whitaker, R.D.: An historical note on the conservation of mass. J. Chem. Educ. **52**(10), 658 (1975)
57. Kamand, F.Z.: Hydraulic friction factors for pipe flow. J. Irrig. Drain. Eng. **114**(2), 311–323 (1988)
58. Loudon, C., McCulloh, K.: Application of the Hagen—Poiseuille equation to fluid feeding through short tubes. Ann. Entomol. Soc. Am. **92**(1), 153–158 (1999)
59. Roldán, C., Campa, F., Altuzarra, O., Amezua, E.: Automatic identification of the inertia and friction of an electromechanical actuator. In: New Advances in Mechanisms, Transmissions and Applications, pp. 409–416. Springer (2014)

60. Tomei, P.: Adaptive PD controller for robot manipulators. IEEE Trans. Robot. Autom. **7**(4), 565–570 (1991)
61. Sun, J., Chen, G., Ko, K.-T., Chan, S., Zukerman, M.: PD-controller: a new active queue management scheme. In: GLOBECOM'03. IEEE Global Telecommunications Conference (IEEE Cat. No. 03CH37489), pp. 3103–3107. IEEE (2003)
62. Wang, J.S., Zhang, Y., Wang, W.: Optimal design of PI/PD controller for non-minimum phase system. Trans. Inst. Meas. Control **28**(1), 27–35 (2006)
63. Kuo, Y.P., Li, T.S.: GA-based fuzzy PI/PD controller for automotive active suspension system. IEEE Trans. Ind. Electron. **46**(6), 1051–1056 (1999)
64. Hoang, N.Q., Kreuzer, E.: Adaptive PD-controller for positioning of a remotely operated vehicle close to an underwater structure: theory and experiments. Control Eng. Pract. **15**(4), 411–419 (2007)
65. Lewis, F.L., Dawson, D.M., Abdallah, C.T.: Robot Manipulator Control: Theory and Practice. CRC Press (2003)
66. Van den Bos, A.: Parameter Estimation for Scientists and Engineers. Wiley (2007)
67. Sontag, E.D., Wang, Y.: On characterizations of the input-to-state stability property. Syst. Control Lett. **24**(5), 351–360 (1995)

Printed in the United States
by Baker & Taylor Publisher Services